GRAPHING CALCULATOR AND EXCEL® SPREADSHEET MANUAL

KARLA NEAL
Louisiana State University

DALE R. BUSKE
St. Cloud State University

FINITE MATHEMATICS FOR BUSINESS, ECONOMICS, LIFE SCIENCES AND SOCIAL SCIENCES

TWELFTH EDITION

Raymond A. Barnett
Merritt College

Michael R. Ziegler
Marquette University

Karl E. Byleen
Marquette University

Prentice Hall
is an imprint of

The author and publisher of this book have used their best efforts in preparing this book. These efforts include the development, research, and testing of the theories and programs to determine their effectiveness. The author and publisher make no warranty of any kind, expressed or implied, with regard to these programs or the documentation contained in this book. The author and publisher shall not be liable in any event for incidental or consequential damages in connection with, or arising out of, the furnishing, performance, or use of these programs.

Reproduced by Pearson Prentice Hall from electronic files supplied by the authors.

ISBN-13: 978-0-321-64541-8
ISBN-10: 0-321-64541-3

1 2 3 4 5 6 BB 14 13 12 11 10

Prentice Hall
is an imprint of

www.pearsonhighered.com

GRAPHING CALCULATOR MANUAL

KARLA NEAL

Louisiana State University

FINITE MATHEMATICS FOR BUSINESS, ECONOMICS, LIFE SCIENCES AND SOCIAL SCIENCES

TWELFTH EDITION

Prentice Hall
is an imprint of

CONTENTS

Chapter 1: Linear Equations and Graphs

For additional help on all aspects of the TI-83 or TI-84 graphing calculator, go to
http://www.prenhall.com/divisions/esm/app/graphing/ti83/

Graphing calculators are particularly efficient tools to use when solving linear equations and inequalities. They can be used to check work as well.

Section 1-1 Linear Equations and Graphs

Example 1: Solving a Linear Equation (page 3)
Solve and check: $8x - 3(x-4) = 3(x-4) + 6$

We can use a calculator to check an answer with work that has been done algebraically by storing the value of the variable. In this problem, the solution was found to be $x = -9$. Type in the number -9 in your calculator, press STO, then select the variable X, and press ENTER. After you have done that, you should type in each side of the equation, using the variable X. If the answer is correct, both sides will have the same value.

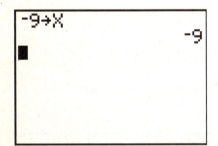

Exercise 1.1.18 (page 11) Solve $-3(4-x) = 5 - (x+1)$

For this example, we will solve the equation graphically. Enter both equations in the Y= menu. You will have to set the Window so that the point of intersection is shown. It's a good idea to start out using the ZOOM Standard window, and then you can adjust from there. Scroll down to ZStandard and press ENTER. The graphs will be drawn.

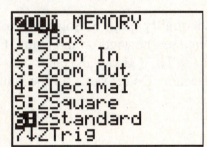
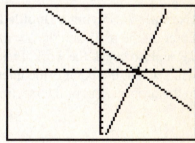

Now you will find the point of intersection, which is the solution to the equation.
Press 2nd TRACE (CALC) and scroll down to 5:intersect, and press ENTER. Move the cursor to near the point of intersection and press ENTER.

1

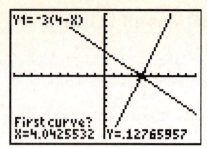
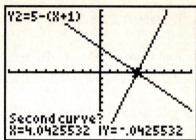

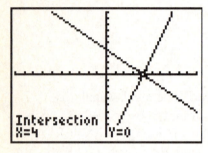 Press ENTER until you see the screen shown.

The solution to the equation is $x = 4$.

Example 6: Solving a Linear Inequality (page 7) Solve and graph: $2(2x+3) < 6(x-2)+10$

You can check the algebraic work graphically. Enter both sides of the inequality into the Y= menu.

You will have to set the window to show the intersection point. Find the point of intersection. You can see that the point of intersection is 4, and the solution is correct.

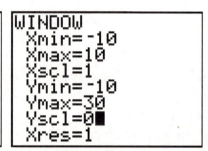
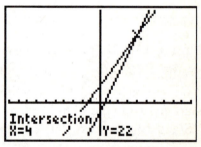

Example 7: Solving a Double Inequality (page 8) Solve and graph $-3 < 2x+3 \le 9$.

Enter all the parts into the Y= menu, set the window and graph. For this one, we will use TRACE to check the work. Press the TRACE key and enter the value 3 and press ENTER. You will see that the point is the intersection. Do the same for the value -3. You can toggle the cursor to the proper point of intersection.

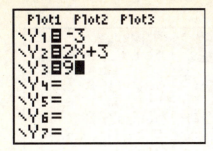
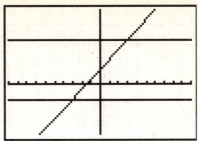
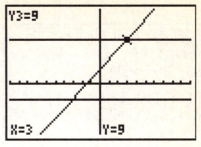

Section 1-2 Graphs and Lines
Example 2 Using a Graphing Calculator (page 15)
Graph $3x - 4y = 12$ and find the intercepts.

Enter $3x / 4 - 3$ in the Y= menu. Use the ZOOM 6: Standard window and graph. To find the y-intercept, use TRACE key and enter x=0.

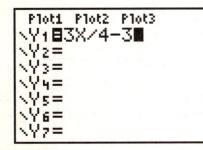
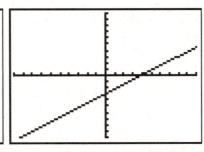
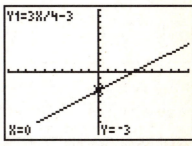

To find the x-intercept, you will use the ZERO function in the CALC menu. Press 2nd TRACE (CALC) and a window will prompt you to enter a left bound. Press ENTER. You will then be prompted to designate a right bound. Move the cursor until it is to the right of the intercept. Press ENTER twice until the ZERO is found.

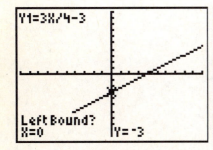
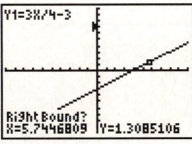
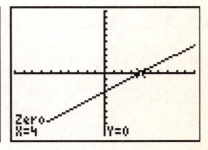

The x-intercept is 4, and the y-intercept is -3.

Section 1-3 Linear Regression

Example 3 Diamond Prices (page 29)
To solve this problem using linear regression requires that the data points be loaded into a list. Press STAT and then EDIT. Put the numbers in each list. Then enter 2nd Y= to go to STAT PLOT.

3

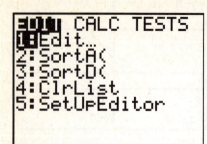
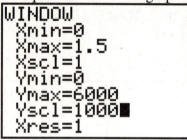

The next window should be configured as shown below on the left. Set the WINDOW to fit the data.

Press GRAPH. Make sure no other equations are in the graph menu.

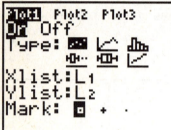
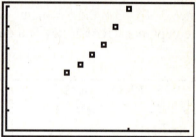
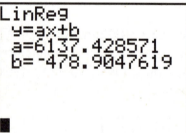

Now go to the STAT CALC menu and select 4:LinReg(ax+b). Press ENTER twice until the screen shows the equation.

4

Chapter 2: Functions and Graphs

Graphing a function is an important skill that is greatly aided with a graphing calculator.

For additional help on all aspects of the TI-83 or TI-84 graphing calculator, go to http://www.prenhall.com/divisions/esm/app/graphing/ti83/

Section 2-1 Functions

The TABLE feature is very useful to evaluate the function at a lot of values very quickly. It is more difficult to use it when the relation being graphed is not a function.

Example 1: Point-by-point plotting (page 44). Sketch the graph of (A) $y = 9 - x^2$.

Open the Y= window and type in the function. You will need to set up the TABLE. Press 2nd WINDOW (TBLSET) and make sure your screen looks like the 2nd window shown below. Then go to the TABLE by entering 2nd GRAPH. Enter the values you want to plot. Use the cursor to scroll up and down if you want to change the X values. You can use the values in the TABLE to help you set a window for graphing. (Make certain that Plot1, Plot2, and Plot3 are off.)

 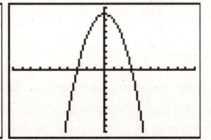

Example 4: Function Evaluation (page 49). If $f(x) = \dfrac{12}{x-2}$ $\quad g(x) = 1 - x^2$ $\quad h(x) = \sqrt{x-1}$

then find $(A)\ f(6)$ $\quad (B)\ g(-2)$ $\quad (C)\ h(-2)$.

There is more than one way to do this problem on the calculator. What we will do here is show a different method for each part.

$(A)\ f(6)$ For this one, we will use the TABLE function. Enter the function $f(x) = \dfrac{12}{x-2}$ in the Y= menu. Put parentheses around the denominator. Once the function is entered, access TABLE and type in 6.

5

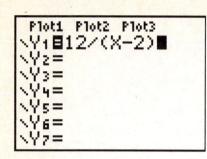 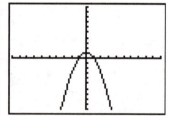

$$f(6)=3$$

(B) $g(-2)$ For this problem, we will use the graph to evaluate the function. Enter the function $g(x)=1-x^2$ into the Y= menu, set the window to Standard by pressing ZOOM 6. The graph will be created.

Now press 2nd CALC and select the first choice (1:Value). Enter the number -2 and press ENTER. You can see that the value of the function at -2 is -3. You can enter another value if you want, but it must be within the window range. For values outside the window range, use the table or reset the window.

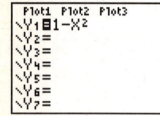 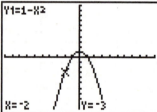

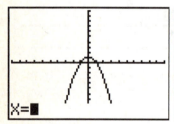

(C) $h(-2)$ We will evaluate this function on the home screen. This is a good method if you need only one value, but is not really useful for multiple values.

On the home screen use the store function (STO) to store the value -2 into the variable X. Press -2 STO X ENTER. Then type in the function using the variable. The calculator will tell you that the function is not defined as a real number at the value x=-2.

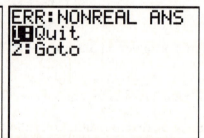

Exercise 2.1.59 (page 55) Find the domain of $g(x) = \sqrt{7-x}$

A graphing calculator can be used to help you check the domain of a function. You cannot necessarily use it exclusively, but it can be helpful. Algebraically, you should determine that the domain is the solution to the inequality $7 - x \geq 0$, which is $x \leq 7$.

Enter the function and graph it.

  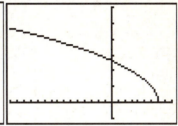

Use the TRACE key and enter the value of 7. Then enter a value greater than 7. You can determine that the domain set as $x \leq 7$ is correct.

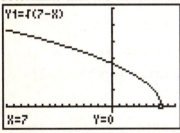 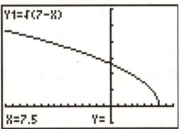

Section 2-2 Elementary Functions: Graphs and Transformations

A graphing calculator is especially useful in seeing graph transformations.

Example 2: Vertical and Horizontal Shifts (page 61). (A) How are the graphs of the functions $y = |x|$, $y = |x| + 4$, $y = |x| - 5$ **related? (B) How are the graphs of** $y = |x|$, $y = |x+4|$, $y = |x-5|$ **related?**

Enter the three functions in the Y= menu. Set the WINDOW to ZOOM Standard and compare the graphs. The absolute value function is found under MATH NUM window. You should access that from the Y= window. Just press ENTER and it will be transferred to the graph window. You can use the TRACE key to note values.

(A) $y = |x|$, $y = |x| + 4$, $y = |x| - 5$

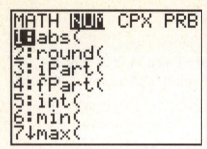 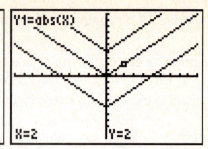

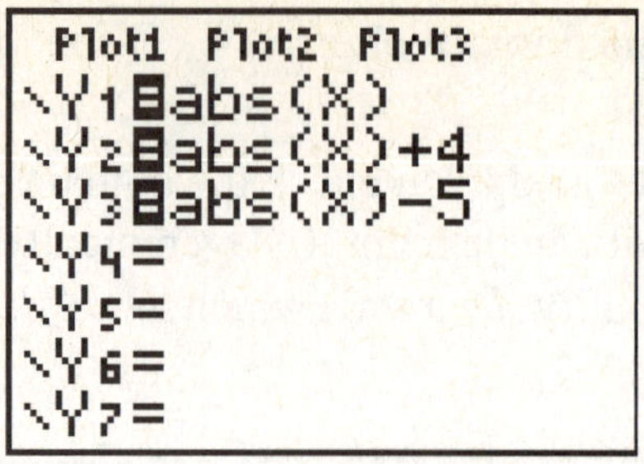

(B) $y = |x|$, $y = |x+4|$, $y = |x-5|$

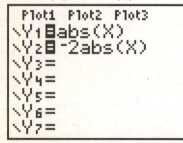 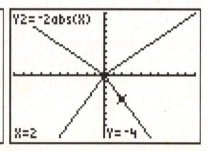

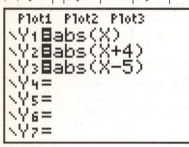 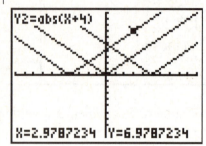

Example 4: Reflections, Stretches, and Shrinks (page 63). (A) How are the graphs of the functions $y = |x|$, $y = 2|x|$, $y = 0.5|x|$ related? (B) How are the graphs of $y = |x|$, $y = -2|x|$ related?

Enter the functions as before. Use the TRACE key to toggle back and forth between the functions so that you can see the function used to create that graph.

(A) $y = |x|$, $y = 2|x|$, $y = 0.5|x|$

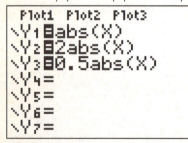 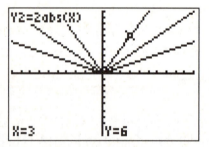

(B) $y = |x|$, $y = -2|x|$

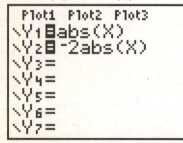 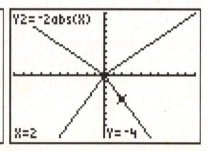

Example 5: Reflections, Stretches, and Shrinks (page 64). How are the graphs of the functions $y = |x|$ and $y = -|x-3|+1$ related?

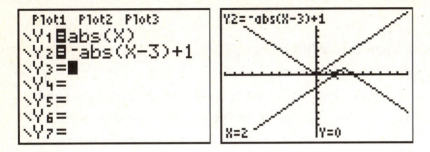

Exercise 2.2.47 (page 67) Graph the function $h(x) = \begin{cases} 2x & \text{if } 0 \le x \le 20 \\ x+20 & \text{if } 20 < x \le 40 \\ 0.5x+40 & \text{if } x > 40 \end{cases}$

This type of function is called a piecewise-defined function. These can be graphed on a graphing calculator, but must be entered carefully. You will need to use the TEST and LOGIC menus which are found by entering 2nd MATH. You will enter each piece of the function using this menu using a + sign between each piece. You will also need to put the graph in DOT mode. You can put the graph in DOT mode directly from the graph editor. Move the cursor all the way to the left and select ENTER until you see the dots. If the pieces of the graph do not meet, then putting the graph in dot mode will indicate the gap. This particular function does not have a gap in the graph.

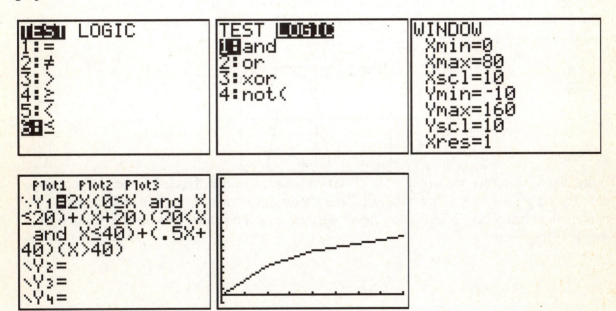

Section 2-3 Quadratic Functions

The example below introduces several ways that a calculator can be used to solve equations and inequalities.

Example 1: Intercepts, Equations, and Inequalities (page 71).
(C) Graph $f(x) = -x^2 + 5x + 3$ in a standard viewing window.
(D) Find the x and y intercepts to four decimal places using trace and the zero command.
(E) Solve the quadratic inequality $-x^2 + 5x + 3 \geq 0$ graphically to four decimal places.
(F) Solve the equation $-x^2 + 5x + 3 = 4$ to four decimal places graphically using intersection.

(C) Enter the function and use ZOOM:6 as the window. Make sure you use the negative sign and not the subtraction sign.

  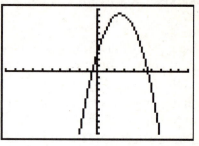

(D) Find the x and y intercepts to four decimal places using trace and the zero command.
The y-intercept is the value of the function when $x=0$. With the graph screen showing, press TRACE. The value at the center of the screen will be shown. In this case, it is for x=0. We see that the value of the y-intercept is 3.

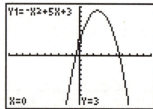

To find the x-intercepts, we will use the ZERO command found in the CALC menu. From the graph screen press 2nd TRACE and select 2:zero. Move the cursor until it is below the first intercept and then press ENTER. Now move the cursor until it is above the intercept and then press ENTER twice.

10

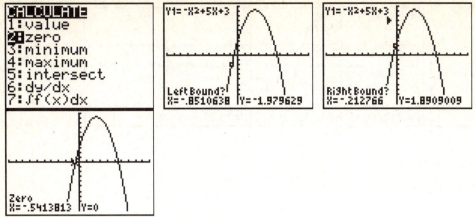

Repeat the process for the other zero.

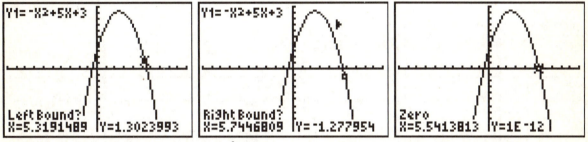

(E) Solve the quadratic inequality $-x^2 + 5x + 3 \geq 0$ graphically to four decimal places.
To solve this inequality, simply use the values found in part (D) to form the interval where the function lies above the x-axis. From this you can determine that $-x^2 + 5x + 3 \geq 0$ over the interval [-0.541, 5.541]

(F) Solve the equation $-x^2 + 5x + 3 = 4$ to four decimal places graphically using intersection.
Enter the function y=4 into the Y= menu and graph.

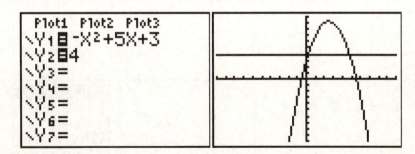

To find the points of intersection, press 2nd TRACE and select 5:intersect. Move the cursor until it is close to one point of intersection. Press ENTER until you see the point of intersection.

11

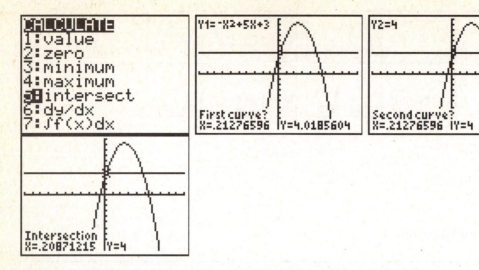

Repeat to find the other point of intersection.

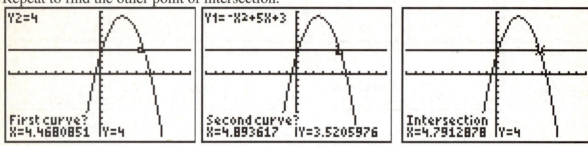

Example 2: Analyzing a Quadratic Function (page 75). Given the quadratic function
$f(x) = 0.5x^2 - 6x + 21$ **(E) Graph the function using a suitable viewing window and (F)**
Find the vertex and the maximum or minimum.

(E) Finding the window can be done several ways. The TABLE is useful, or you can use
TRACE. The STANDARD window is a good starting point. For this function, this will be
sufficient to use for finding the vertex. Once the vertex can be seen, you can find the coordinates
of the vertex using the CALC menu. For this function, we will select 3:minimum.

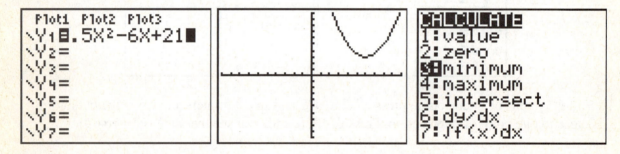

Move the cursor until it is close to the minimum but still to the left. Press ENTER and then move
to a right bound. Press ENTER until you see the point. The minimum is 3 and the vertex is (6,3).

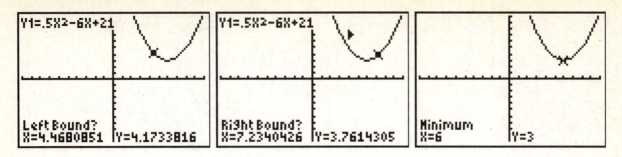

Example 5: Outboard Motors (page 79).

To solve this problem using quadratic regression requires that the data points be loaded into a list.
Press STAT and then EDIT. Put the numbers in each list. Then enter 2nd Y= to go to the STAT PLOT.

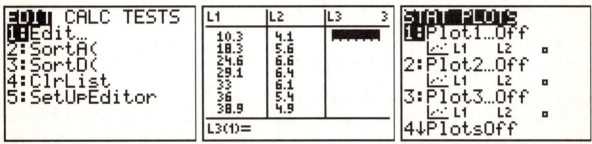

The next window should be configured as shown below on the left. Set the WINDOW to fit the data.
Press GRAPH. Make sure no other equations are in the graph menu.

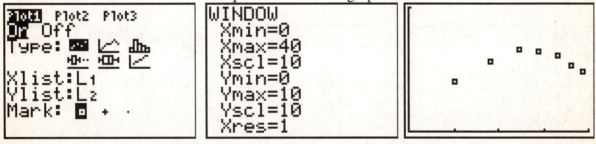

Now go to the STAT CALC menu and select 5:QuadReg. Press ENTER twice until the screen shows the equation. Graph the function and use trace to find other points. To have the regression equation automatically entered into the Y= menu, from the Y= menu, go to the VARS menu, select 5:Statistics, then select EQ and 1:RegEQ and press ENTER.

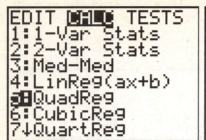 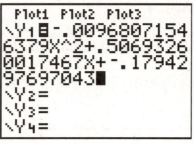 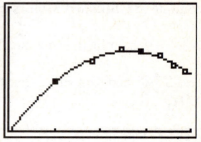

Press GRAPH and the points and the equation are graphed. You can look at the equation in the Y= menu.

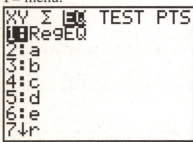

Section 2-4 Polynomial and Rational Functions

Example 1 Estimating the Weight of a Fish (page 87)
To solve this problem, cubic regression is used, much like quadratic regression as used in section 2-3.

Press STAT and then EDIT. Put the numbers in each list. Then enter 2nd Y= to go to the STAT PLOT.

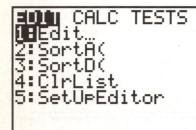 

The next window should be configured as shown below on the left. Set the WINDOW to fit the data.
Press GRAPH. Make sure no other equations are in the graph menu.

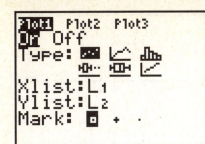

Now go to the STAT CALC menu and select 6:CubicReg. Press ENTER twice until the screen shows the equation. Graph the function and use trace to find other points. To have the regression equation automatically entered into the Y= menu, from the Y= menu, go to the VARS menu, select 5:Statistics, then select EQ and 1:RegEQ and press ENTER. Graph the function.

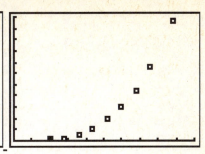

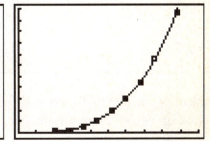

Now use the TABLE to estimate the weights at lengths of 39, 40, 41, 42, and 43 inches.

X	Y1
39	358.28
40	389.72
41	423.02
42	458.26
43	495.48

X=

Example 2: Graphing Rational Functions (page 88) Given the rational function

$f(x) = \dfrac{3x}{x^2 - 4}$ **(E) Using the information from (A)-(D) and additional points as necessary, sketch the graph of f for $-7 \le x \le 7$.**

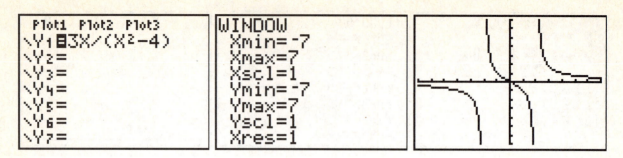

Exercise 2.4.60 (page 94) Minimum average cost. (D) Graph the average cost function $\overline{C}$ on a graphing calculator and use an appropriate command to find the daily production level (to the nearest integer) at which the average cost per player is at a minimum. What is the minimum average cost to the nearest cent?

The function $\overline{C}(x) = \dfrac{x^2 + 2x + 2000}{x}$ is graphed and the minimum is found using the minimum function.

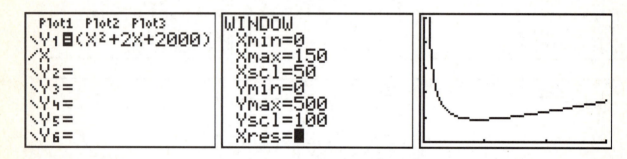

Select the minimum function by pressing 2nd TRACE:3. Move the cursor to a value of x that is less than where the minimum occurs, then press ENTER. Repeat by moving the x to the right of the minimum value.

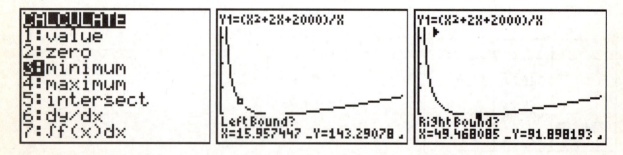

Press ENTER twice so that the minimum is shown. Rounding as designated, it is shown that the minimum is at 45 units; $91.44 per player.

16

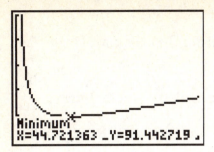

Section 2-5 Exponential Functions

Example 1: Graphing Exponential Functions (page 97) Sketch the graph of
$y = \left(\frac{1}{2}\right)4^x, -2 \le x \le 2.$

Enter the function into the Y= menu and set the window. To determine the values for the window, use the table.
You can see the values of the function for the given interval of x. Set the window and then graph.

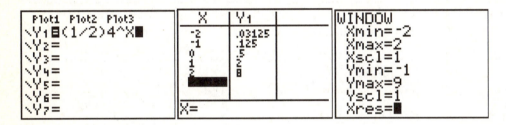

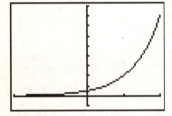

Example 2: Exponential Growth (page 99). For $N = N_0 e^{1.386t}$ **, find the number of bacteria present in (A) 0.6 hour and (B) in 3.5 hours if** $N_0 = 25$.

This can be done using the TABLE function. Enter the function into the graph Y= menu. Go to the table and enter the values 0.6 and 3.5. It is not necessary to graph the function to evaluate it using the TABLE, so you do not need to set a window.

There are 57 bacteria present when x=0.6 and 3197 bacteria present when x=3.5.

You can use the calculator to approximate an exponential curve using the exponential regression function.

Example 4: Depreciation (page 101). Enter the data into the calculator and find the exponential function that fits the data.

Press STAT and then select 1:Edit. Enter the data values into the lists L1 and L2. Now go to the STAT: CALC menu and scroll down to 0: ExpReg

Press ENTER twice until the screen below is seen. You can graph the data points and the exponential regression function. To graph the data points, press 2nd Y= which is the STAT PLOT menu. Press ENTER and access the screen. It should be set up as shown.

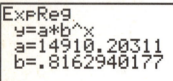

Set the window. To have the regression equation automatically entered into the Y= menu, from the Y= menu, go to the VARS menu, select 5:Statistics, then select EQ and 1:RegEQ and press ENTER.

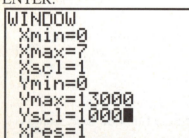

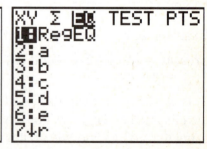

Now press graph and the two graphs are plotted.

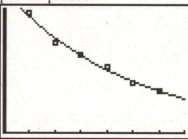

18

Section 2-6 Logarithmic Functions

Your calculator will evaluate logarithms of base 10 and base *e*. These are the keys LOG and LN.

Example 7: Calculator Evaluation of Logarithms (page 111). Use a calculator to evaluate each to six decimal places: (A) log 3,184   (B) ln 0.000349   (C) log (-3.24)

These can all be evaluated on the home screen. Make sure you close the parentheses. After you press ENTER for log (-3.24), you will get the error message shown below since that value is not defined.

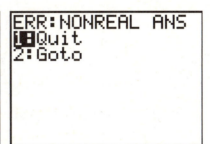

Logarithmic equations can be solved graphically.

Example 8: Solving log$_b$x=y for x. (page 111). Find x to four decimals for log x = -2.315.

Enter two functions into the Y= menu. Set a window, graph, and find the point of intersection.

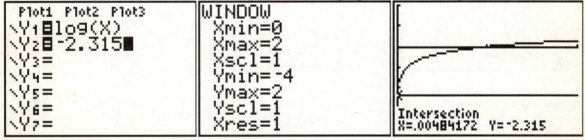

Example 9: Solving Exponential Equations (page 111). Solve for *x* to four decimal places: (A) $10^x = 2$.

This equation can be solved graphically by entering both functions and finding the point of intersection. The intersect function is accessed by pressing 2nd TRACE:5.

19

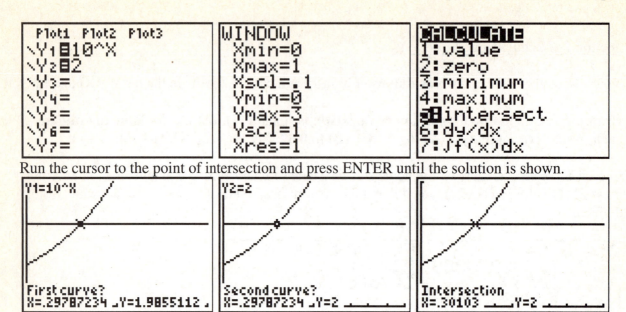

Run the cursor to the point of intersection and press ENTER until the solution is shown.

Example 11: Home Ownership Rates (page 114). Use logarithmic regression to find the best model in the form y=a+b ln x.

As with Example 4 on page 101, we enter the data into the list, find the regression equation, enter the regression equation into the Y= menu and graph the data and the equation.
Let x=0 represent the year 1900.

Once you have set the window and graphed the function, use the VALUE function (2nd CALC 1) to predict that home ownership in 2015 will be 69.4%.

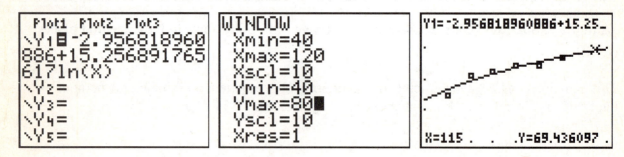

Chapter 3: Mathematics of Finance

For additional help on all aspects of the TI-83 or TI-84 graphing calculator, go to
http://www.prenhall.com/divisions/esm/app/graphing/ti83/

Section 3-1 Simple Interest

Example 2: Present Value of an Investment (page 128). If you want to earn an annual rate of 10% on your investments, how much (to the nearest cent) should you pay for a note that will be worth $5000 in 9 months?

The equation is $A = P(1+rt)$, and we have $5000 = P(1+0.1(0.75)) \Rightarrow P = \dfrac{5000}{(1+0.1(0.75))}$.

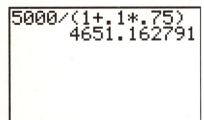

Make certain that you put parentheses around the denominator.

Exercise 29: (page 131) A=$14,560; P=$13,000; t=4 months; r = ?

The formula is $A = P(1+rt)$, and we have $r = \dfrac{A-P}{Pt} = \dfrac{14560-13000}{13000\left(\dfrac{4}{12}\right)}$.

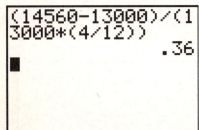

Pay attention to the parentheses around the numerator and the denominator.

Section 3-2 Compound and Continuous Interest

Example 2: Compounding Daily and Continuously (page 136). What amount will an account have after 2 years if $5000 is invested at an annual rate of 8% (A) compounded daily? (B) compounded continuously?

(A) Using the compound interest formula $A = P\left(1+\dfrac{r}{m}\right)^{mt}$, we have $A = 5000\left(1+\dfrac{.08}{365}\right)^{(365\cdot2)}$. Be

careful to place parentheses in the proper places, especially the exponent.

21

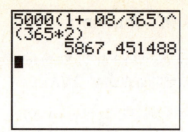

(B) Use the continuous compound interest formula $A = Pe^{rt}$. This is an easier formula to enter into the calculator. Don't forget the parentheses in the exponent. $A = 5000e^{(0.08 \cdot 2)}$.

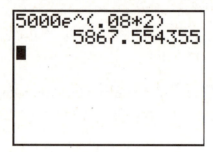

Example 5: Computing Growth Time (page 139). How long will it take $10,000 to grow to $12,000 if it is invested at 9% compounded monthly?

We are solving the equation $12000 = 10000\left(1 + \dfrac{.09}{12}\right)^n$. We will solve this problem graphically.

Enter both sides of the equation into the Y= menu. Set a proper window that will include the point of intersection.
Graph and find the point of intersection. Intersection is found by pressing 2nd TRACE 5. Move the cursor near the point of intersection and press ENTER until you see the screen shown below. We can see that it will take about 25 months.

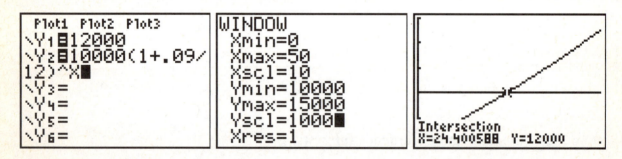

There is another solution method using the Equation Solver. This is found under the MATH menu.
Press ENTER and the Equation Solver will come up. Enter the equation using the ALPHA key.

```
MATH NUM CPX PRB          EQUATION SOLVER          EQUATION SOLVER
4↑³√(                     eqn:0=                   eqn:0=A-P(1+I)^N
5: ×√                                              ▮
6:fMin(
7:fMax(
8:nDeriv(
9:fnInt(
0BSolver…
```

Press ENTER and then put in the values. Leave the value for N blank and then hit ENTER.

```
A-P(1+I)^N=0              A-P(1+I)^N=0
 A=12000                   A=12000
 P=10000                   P=10000
 I=.0075                   I=.0075
 N=▮                     ▪N=24.400588158…
 bound={-1E99,1…           bound={-1E99,1…
                         ▪left-rt=0
```

Section 3-3 Future Value of an Annuity; Sinking Funds

Example 4: Approximating and Interest Rate (page 151). Find the interest rate, *i*, if FV=$160,000, PMT=$100, and *n*=360.

This data results in the equation $160,000 = 100\dfrac{(1+i)^{360}-1}{i}$. This equation can be solved using graphs and the intersect function. It can also be solved using the equation solver which was demonstrated in section 3-2, example 5.

However, there is another method that can be used with the TI-84. It is the TVM Solver. Press APPS and select 1:Finance. The TVM Solver is the first choice. Enter the values shown, leaving the value for I% blank. Press ALPHA and ENTER to solve for I. You may have to put a 0 in for I% until the other values are entered. Then go back and clear the 0.

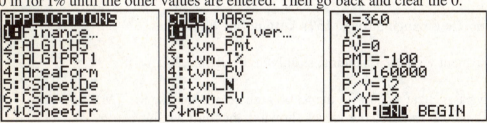

```
APPLICATIONS            CALC VARS               N=360
1BFinance…              1BTVM Solver…           I%=
2:ALG1CH5               2:tvm_Pmt               PV=0
3:ALG1PRT1              3:tvm_I%                 PMT=-100
4:AreaForm              4:tvm_PV                 FV=160000
5:CSheetDe              5:tvm_N                  P/Y=12
6:CSheetEs              6:tvm_FV                 C/Y=12
7↓CSheetFr              7↓npv(                   PMT:END BEGIN
```

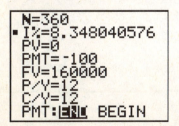

```
 N=360
▪I%=8.348040576
 PV=0
 PMT=-100
 FV=160000
 P/Y=12
 C/Y=12
 PMT:END BEGIN
```

Section 3-4 Present Value of an Annuity; Amortization

Example 2: Retirement Planning (page 155).

What we are trying to do here is to find out how much money to deposit annually for 25 years to receive 20 payments of $25,000 if the interest rate is 6.5% compounded annually.

Find PV with PMT= $25,000, i=0.065, and n=20. Access the TVM Solver as demonstrated in Example 4 from section 3-3. Enter the values and then press ALPHA and ENTER. The amount that will have to be in the account is $275462.68

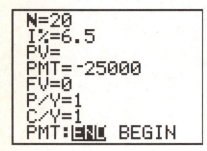  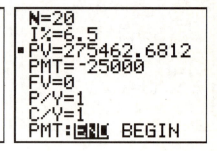

Now that we have the amount that will have to be in the account in 25 years, we will find the amount of 25 annual deposits that will yield that amount. Enter the values as shown and press ALPHA and ENTER.

We now have that an annual payment of $4677.76 for 25 years, will provide $25000 a year for 20 years.

Example 6: Automobile Financing (page 159) Automobile Financing

To determine the payment with 0% financing, simply divide the 25,200 by 48.

To find the monthly payment at 4.5% with the $3,000 rebate, we can use the TMV Solver as demonstrated in Example 2. Entering the values shown in the third window,

Now, move the cursor back to PMT and press Clear. You can evaluate the payment by pressing ALPHA and ENTER.

Chapter 4: Systems of Linear Equations; Matrices

For additional help on all aspects of the TI-83 or TI-84 graphing calculator, go to
http://www.prenhall.com/divisions/esm/app/graphing/ti83/

Section 4-1 Review: Systems of Linear Equations in Two Variables

Example 3: Solving a System Using a Graphing Calculator (page 172).

To solve a system by using a graphing calculator, you will have to put both equations into the Y= menu. You will need to solve each equation for *y*. Then you need to set an appropriate window. If you don't have an idea of the window to use, you can try the ZOOM 6:ZStandard window. Or, you can use the TABLE to locate values of *x* where the two equations are close in value. Once the window is established, you will then press 2nd TRACE (CALC) 5:Intersect. Move the cursor until it is close to the point of intersection and press ENTER until the intersection point is shown.

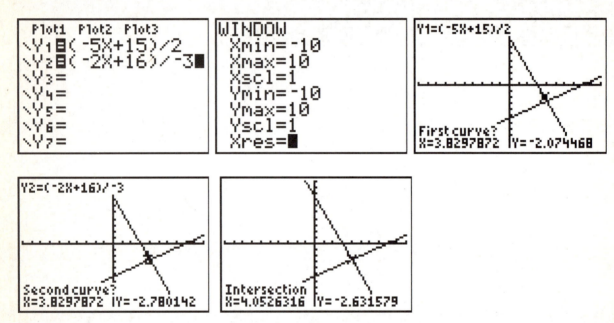

Section 4-2 Systems of Linear Equations and Augmented Matrices

Solving systems of equations can often be solved using augmented matrices. You can command the calculator to perform row operations.

Example 1: Solving a System Using Augmented Matrix Methods (page 185).

To access the Matrix menu, press 2nd and x^{-1}. You will select one of the matrices to enter the values. For this example, we will use matrix [A]. You will have to move the cursor to EDIT to enter the values. The matrix will appear blank unless it has already been edited. Change the dimensions of the matrix to 2 X 3. Move the cursor over the numbers 1 X 1 and change it to 2 X 3. Use the down cursor to move into the matrix and enter the values. As you enter values and press ENTER, the cursor will move to the next space.

26

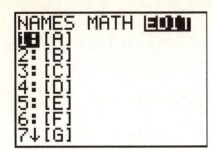
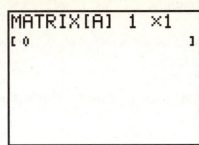
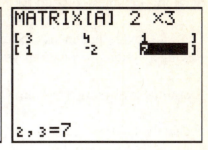

To perform row operations, you first return to the home screen and call up the matrix [A]. Access the MATRIX menu again and press 1. Press ENTER and the matrix name will appear on the home screen. Press ENTER again and you will see the matrix [A]. To perform a row operation, access the matrix menu, move to MATH and scroll down to the row operations.

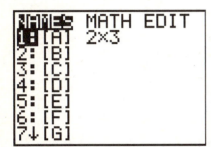
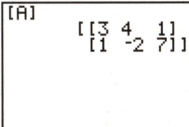

To swap two rows, press **C:rowSwap(** and then enter the name of the matrix, and the rows to be swapped. Then you need to store this new matrix in matrix [A] by pressing STO and entering the matrix name. Then press ENTER. To multiply row 1 by -3, and then add it to row 2 to create a new row 2, you will use the ***row+(** command. These commands can continue until the matrix is in row echelon form.

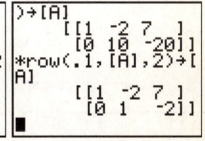

We can see the solution $x_1=3$ and $x_2=-2$.

```
A]
        [[1  -2 7 ]
         [0  1  -2]]
*row+(2,[A],2,1)

        [[1  0  3 ]
         [0  1  -2]]
■
```

27

Section 4-3 Gauss-Jordan Elimination

The TI-84 will put a matrix in row echelon form and reduced row echelon form. This makes solving a system of equations very easy.

Example 2: Solving a System Using Gauss-Jordan Elimination (page 194).

Create the matrix for the system. You will notice that all of the data will not be visible for larger matrices, but you will be able to see this one on the home screen. Store it in [A].

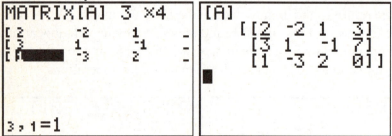

Go into the Matrix menu and select **rref** in order to put the matrix in reduced row echelon form. Notice that one of the entries is a decimal. In order to view this element, scroll to the right. The last number in the second row is 0.

Example 3: Solving a System Using Gauss-Jordan Elimination (page 196).

How can you tell when a system has to no solution using the calculator? You will go through the same steps as in the last example.

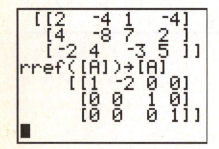

The last row is equivalent to:

$$0x_1 + 0x_2 + 0x_3 = 1.$$

This is obviously a false statement, so we know that the system has no solution.

Example 4: Solving a System Using Gauss-Jordan Elimination (page 197).

How can we tell when a system has infinitely many solutions? This example will demonstrate the way to do this. Enter the matrix as before and put it in reduced row echelon form "rref".

```
[[3    6   -9  15]
 [2    4   -6  10]
 [-2  -3   4   -6]]
rref([A])→[A]
  [[1   0   1   -3]
   [0   1  -2   4 ]
   [0   0   0   0 ]]
■
```

The resulting matrix yields the system
$$x_1 \quad + x_3 = -3$$
$$x_2 - 2x_3 = 4$$
From this the solution to the system can be found.

Section 4-4 Matrices: Basic Operations

The basic operations of addition, subtraction, scalar multiplication, and matrix products can be performed on the calculator.

Example 1: Matrix Addition (page 206)

(B) Perform the addition: $\begin{bmatrix} 2 & -3 & 0 \\ 1 & 2 & -5 \end{bmatrix} + \begin{bmatrix} 3 & 1 & 2 \\ -3 & 2 & 5 \end{bmatrix}$ by entering each matrix. Then enter the name of each matrix with the addition operation.

```
[A]
   [[2 -3 0 ]
    [1 2  -5]]
[B]
   [[3  1 2]
    [-3 2 5]]
```

```
[A]+[B]
   [[5  -2 2]
 ■  [-2 4  0]]
```

The same procedure is used for subtraction of matrices.

Example 4: Multiplication of a Matrix by a Number (page 207).

Enter the matrix and then multiply by -2.

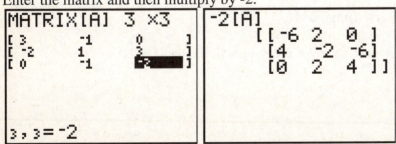

```
MATRIX[A] 3 ×3
[ 3    -1    0    ]
[ -2   1     3    ]
[ 0    -1    -2   ]



3,3=-2
```

```
-2[A]
   [[-6 2  0 ]
    [4  -2 -6]
    [0  2  4 ]]
```

Example 8: Matrix Multiplication (page 210).

(A) To multiply two matrices, they must have the proper dimensions. If they do not, the calculator will show an error message.

Enter the two matrices and then multiply them together.

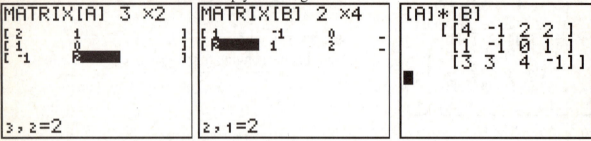

Section 4-5 Inverse of a Square Matrix

If a matrix has an inverse, it can be found on the calculator. The process can be quite long when done by hand, so the calculator can be a real time saver.

Example 2: Finding the Inverse of a Matrix (page 221).

Find the inverse of the matrix.

$$M = \begin{bmatrix} 1 & -1 & 1 \\ 0 & 2 & -1 \\ 2 & 3 & 0 \end{bmatrix}$$

Enter the matrix into the calculator. Call up the matrix on the home screen and then press the x^{-1} key on the calculator. The inverse of the matrix will be displayed.

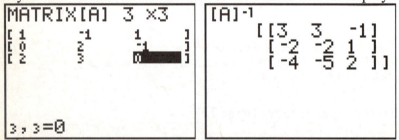

Example 4: Finding a Matrix Inverse (page 224).

If the calculator does not have an inverse, an error message will be displayed.

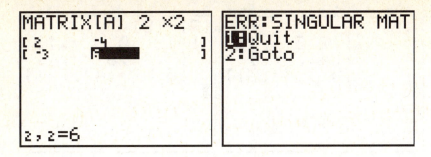

Section 4-6 Matrix Equations and Systems of Linear Equations

The methods shown in this section are primarily for calculations done by hand. The ability of a graphing calculator to put a matrix into reduced form can also be used, but an inverse matrix can also be used.

Some of the homework problems can be done using the graphing calculator.

Exercise 13 (page 234) Find x_1 and x_2.

$$\begin{bmatrix} 1 & -1 \\ 1 & -2 \end{bmatrix} \begin{bmatrix} x_1 \\ x_2 \end{bmatrix} = \begin{bmatrix} 5 \\ 7 \end{bmatrix}$$ This uses the property $AX = B \rightarrow X = A^{-1}B$. We will use

$$A = \begin{bmatrix} 1 & -1 \\ 1 & -2 \end{bmatrix}, B = \begin{bmatrix} 5 \\ 7 \end{bmatrix}.$$

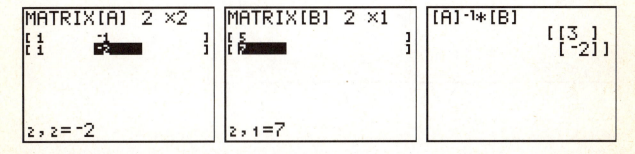

So $x_1 = 3$ and $x_2 = -2$.

Section 4-7 Leontief Input-Output Analysis

Example 1: Input-Output (Analysis page 241)

The rather involved use of matrices in this problem can be done quickly with the calculator. You will have to store the matrices. Store the identity matrix, I, in [A]. Store matrix M in [B].

31

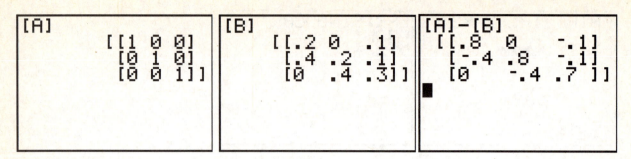

Create matrix [C] by the subtraction [A]-[B] (I-M). Then create the inverse matrix $[C]^{-1}$ ($[I-M]^{-1}$) and store it in [D]. Create matrix [E]. The solution is found by multiplying [D][E].

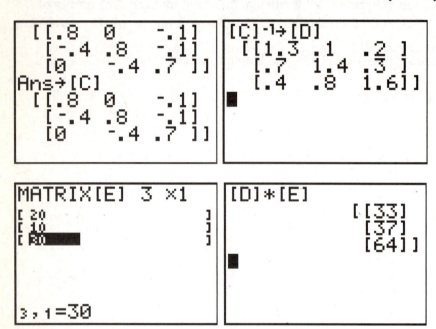

Chapter 5: Linear Inequalities and Linear Programming

For additional help on all aspects of the TI-83 or TI-84 graphing calculator, go to http://www.prenhall.com/divisions/esm/app/graphing/ti83/ There is a section on using the Inequality Graphing Application.

Section 5-1 Inequalities in Two Variables

The TI-84 will shade an area above or below a line.

Example 1: Graphing a Linear Inequality (page 253). Graph $2x - 3y \le 6$.

First you must solve the inequality for y in order to put it into the Y= menu. $\dfrac{2x}{3} - 2 \le y$.

Enter this into the Y= menu. Set the window to ZOOM 6, and select GRAPH.

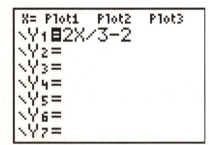 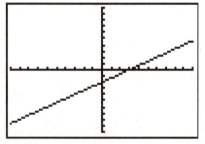

After inserting the test point (0,0), it is determined that the point is in the solution set. In order to shade the area above the line, go back to the Y= menu and move the cursor to the left of Y_1. The line will start flashing. Hit the ENTER key until you see the screen below and then hit GRAPH. The triangle indicates that the area above the line will be shaded. The shaded area is the solution to the inequality.

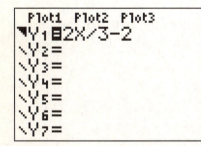 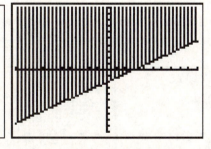

Section 5-2 Systems of Linear Inequalities in Two Variables

The TI-84 has a built in application that will graph and shade the solution area of a system of inequalities. Select the APPS key and scroll to the application **Inequalz**. Open the application and you will be taken to the graph editor. Move the cursor over the = sign. The symbols in the menu at the bottom are accessed via F1-F5 by pressing ALPHA and then the number you wish.

33

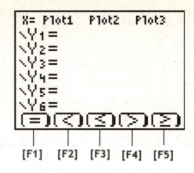

We will now solve a system using this application.

Example 1: Solving a System of Linear Equations Graphically (page 259).
Solve the system

$$x + y \geq 6$$
$$2x - y \geq 0$$

Solve each for y.

$$y \geq 6 - x$$
$$y \leq 2x$$

Enter each inequality. Move the cursor over the = sign and enter ALPHA F5 to select $\geq$ for Y1 and ALPHA F3 to select $\leq$ for Y2. Then press GRAPH. The appropriate areas are shaded. Press ALPHA F1 to access the shades menu.

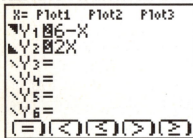
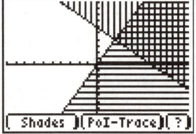
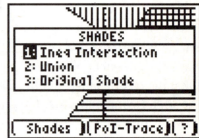

Select the first choice 1:Ineq Intersection to show only the intersection of the two. To find the point of intersection, press ALPHA F3. The intersection point is shown.

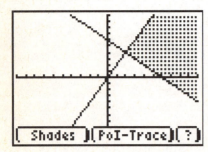
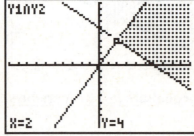

Example 2: Solving a System of Linear Inequalities Graphically (page 260).

We can solve the system of inequalities as in the last example. Take each inequality and simplify. Enter each inequality into the Y= menu, except for $x \geq 0$. To enter that inequality, move the cursor to the top left of the screen where you see the X=. Press ENTER and then put $x \geq 0$ into that menu.

$$2x + y \leq 22 \qquad y \leq 22 - 2x$$
$$x + y \leq 13 \qquad y \leq 13 - x$$
$$2x + 5y \leq 50 \quad \rightarrow \quad y \leq 10 - 2x/5$$
$$x \geq 0 \qquad\qquad x \geq 0$$
$$y \geq 0 \qquad\qquad y \geq 0$$

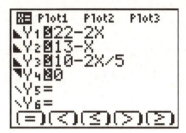 

Set the window and then press GRAPH. Now you will shade the intersection. Press ALPHA F1 and select Intersection.

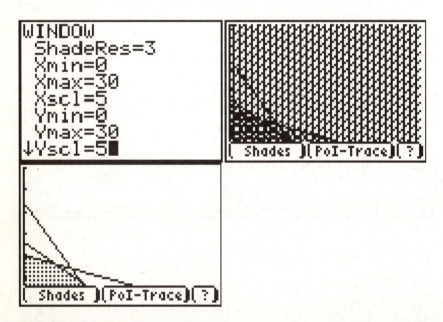

Now select ALPHA F3 and use the up and down cursor as well as the left and right cursor to move to the points of intersection. Notice that the names of the graphs are shown in the upper left of the screen.

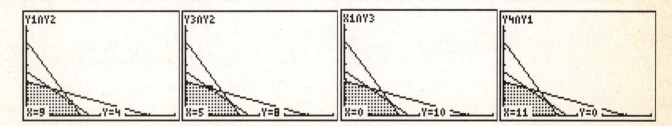

Section 5-3 Linear Programming in Two Dimensions: Geometric Approach

Example 2: Solving a Linear Programming Problem (page 270).
(A) To solve this example, we will use the same procedure as we did in Example 2, section 5-2. You will use the application that graphs inequalities. Select the APPS key and scroll to the application **Inequalz**.

Solve each inequality for y and then enter into the Y=menu.

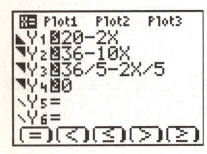 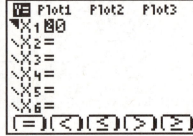 

Graph the inequalities, then select the intersection by pressing ALPHA F1 and selecting option 1. To find the points of intersection, press ALPHA F3 and move the cursor to the intersection points.

 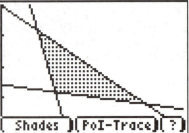 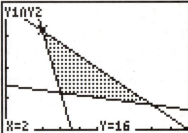

You can go to the home screen to evaluate the constraint equation.

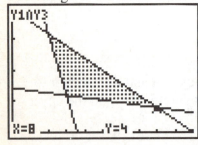 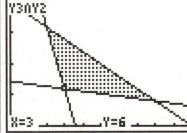 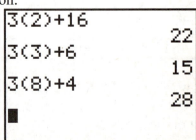

Chapter 6: Linear Programming – Simplex Method

For additional help on all aspects of the TI-83 or TI-84 graphing calculator, go to
http://www.prenhall.com/divisions/esm/app/graphing/ti83/

Section 6-1 A Geometric Introduction to the Simplex Method

No problems with the calculator in this section.

Section 6-2 The Simplex Method: Maximization with Problem Constraints of the Form ≤

Example 1: Using the Simplex Method (page 294).

Enter the simplex tableau in matrix [A]. Now perform the row operations to create the matrix that shows the optimal solution. The row operations are found in the matrix MATH menu.

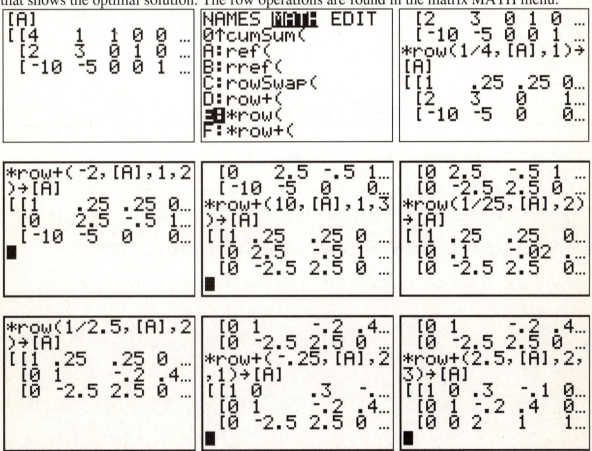

Once you have performed all of the row operations, you can use the cursor to scroll to the right to see the solutions.

Section 6-3 Dual Problem: Minimization with Problem Constraints of the Form ≥

Example 1: Forming the Dual Problem (page 303)

Enter matrix [A] and then create the transpose of the matrix.

Example 2: Solving a Minimization Problem (page 306).

Once you have created the tableau using the transpose matrix from Example 1, you follow the same procedures as in section 6-2, example 1, shown on the previous page.

Section 6-4 A Maximization and Minimization with Mixed Problem Constraints

No new methods of solutions are done in this section. You can use the methods covered in sections 6-2 and 6-3

Chapter 7: Logic, Sets, and Counting

For additional help on all aspects of the TI-83 or TI-84 graphing calculator, go to
http://www.prenhall.com/divisions/esm/app/graphing/ti83/

Section 7-1 Logic

To access the Logic menu, press 2nd Math, which is the TEST menu, then use the cursor to select the LOGIC menu. There are four logic operations that you can use. The operations **and, or,** and **xor** (exclusive or) return a value of 1 if an expression is true or 0 if an expression is false. The operation **not** returns 1 if the value is 0.

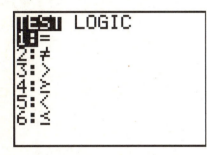

Example 5: Constructing a Truth Table (page 339) Construct the truth table for
$p \wedge \neg (p \vee q).$

Enter the values for p and q in L1 and L2. Press STAT and EDIT. Then you can put the values into the lists.

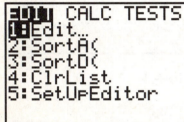 

Now you will use the logic menu to put values into other lists. Press 2nd STAT (LIST) to access the names of the lists that are stored in the calculator. Then use the LOGIC menu as instructed above, and store the values in the given lists as shown.

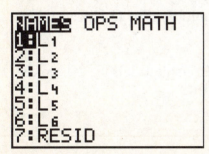 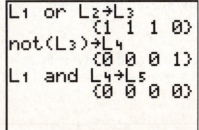

39

Example 7: Verifying a Logical Equivalence (page 341) Show that $\neg(p \vee q) \equiv \neg p \wedge \neg q$.

Enter the values for p and q in L1 and L2 as shown in the previous example. Store $p \vee q$ in L3 and store $\neg(p \vee q)$ **in** L4. Store $\neg p \wedge \neg q$ in L5. You can see that L4 and L5 are equivalent.

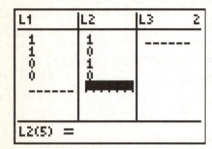 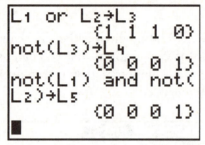

Section 7-2 Sets and **Section 7-3**

No problems from 7-2 or 7-3 can be done on the calculator.

Section 7-4 Permutations and Combinations

Example 1: Computing Factorials (page 357)

(A) $5!$ (B) $\dfrac{7!}{6!}$ (C) $\dfrac{8!}{5!}$ (D) $\dfrac{52!}{5!47!}$

To calculate a factorial, enter the number, then MATH and scroll over to the PRB (probability) menu. The factorial function is selection 4. Now you can calculate the factorial expressions.

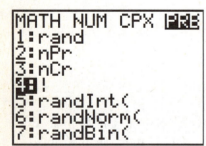 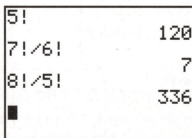

Example 3: Permutations (page 378) Find the number of permutations of 13 objects taken 8 at a time.

The permutation function is found on the probability menu as in Example 1. It is the 2nd choice on the menu. You must enter the 13, then access **nPr**, 8 and ENTER.

40

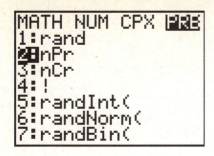 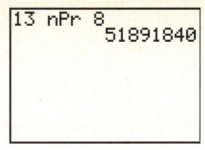

Example 4: Permutations and Combinations (page 361)

(A) This part of the problem uses a permutation. Enter the number 10 on the home screen, and then go to the probability menu and select $2:n\Pr$, then enter 3 and ENTER.

(B) This part of the problem uses a different function called a combination. The function is accessed from the Probability menu as is the permutation. Enter the number 10, and then go to the probability menu and select $3:nCr$, then enter 3 and ENTER.

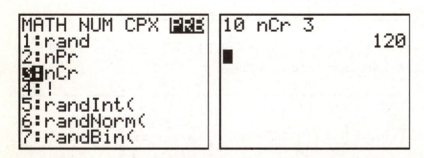

Example 5: Combinations (page 362). Find the number of combinations of 13 objects taken 8 at a time.

The combination function is found on the probability menu as in the last example. It is the 3rd choice on the menu. You must enter the 13, then access **nCr** and then enter the 8.

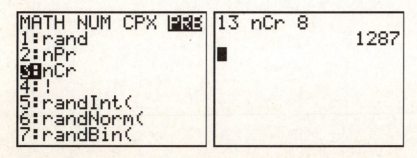

Chapter 8: Probability

For additional help on all aspects of the TI-83 or TI-84 graphing calculator, go to
http://www.prenhall.com/divisions/esm/app/graphing/ti83/

Section 8-1 Sample Spaces, Events, and Probability

Example 6: Simulation and Empirical Probabilities (page 381) Simulate 100 rolls of two dice with the random number generator on a graphing calculator.

The random number generator is found in the MATH PRB menu. This will generate integers. What we are going to do is have the calculator generate 100 random integers between 1 and 6, and this will be done twice. These two random numbers will be added together and stored in a list. We will use L1. Then we will create a histogram of the values found. Note that the same numbers will not be generated by your calculator that are generated by the example shown in the text. Recall that the list names are found by pressing 2nd STAT.

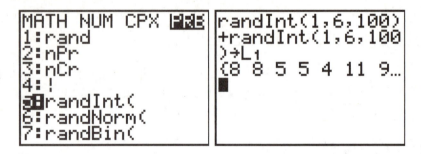

Now we will create a histogram. Press 2nd Y= to call up the STAT PLOT menu. Press ENTER and turn the plot on. Select the histogram icon. Set up a window.

Press 2nd ZOOM (FORMAT) and select GridOn. Then press GRAPH. To see the values, use the TRACE key.

42

 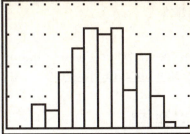 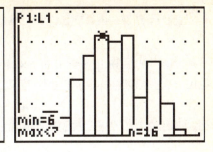

Section 8-2 Union, Intersection, and Complement of Events; Odds

Exercise 69, page (396). Use a graphing calculator to simulate 50 repetitions of rolling a pair of dice and recording their sum, and find the empirical probability of rolling a 7 or 8.

We will use the same commands as in the previous section. The only change will be that the number of repetitions will be changed to 200. Then the histogram will be created, and the TRACE key will be used to find the number of times that a sum of 7 occurred and the number of times that a sum of 8 occurred.

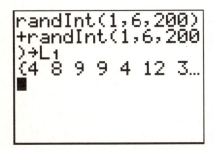  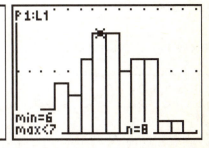

Both sums occurred 8 times, so the probability is (8+8)/50=0.32.

Section 8-3 Conditional Probability, Intersection, and Independence

No problems in this section introduce any new concepts. For directions on how to use the calculator to find factorials, combinations and permutations, see sections 7-3, and 7-4.

Section 8-4 Bayes' Formula

No problems in this section introduce any new concepts. For directions on how to use the calculator to find factorials, combinations and permutations, see sections 7-3, and 7-4.

Section 8-5 Random Variable, Probability Distribution, and Expected Value

No problems in this section introduce any new concepts. For directions on how to use the calculator to generate a table of random values, see the examples in section 8-1 and 8-2.

Chapter 9: Markov Chains

For additional help on all aspects of the TI-83 or TI-84 graphing calculator, go to http://www.prenhall.com/divisions/esm/app/graphing/ti83/

Section 9-1 Properties of Markov Chains

Example 3: Using a Graphing Calculator and P^k to Compute S_k (page 438).

Store the matrices. Place P in matrix [A] and S_0 in matrix [B]. You can use the home screen to check your values.

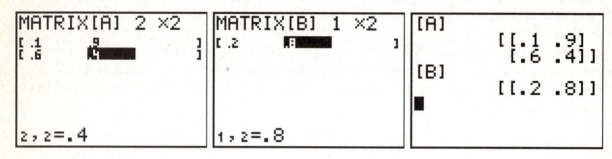

Now calculate $S_8 = S_0 P^8$.

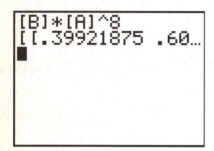

Section 9-2 Regular Markov Chains

Example 5: Approximating the Stationary Matrix (page 448). Use a graphing calculator to find a stationary matrix.

Put the transition matrix into the calculator. Set the decimal display to four places using the MODE menu. Compute powers of the matrix until all the rows are the same.

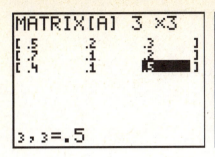  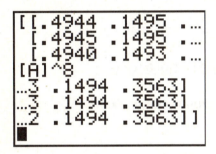

Scroll to the right to see the last column.

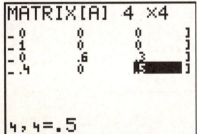

We can conclude that the transition matrix is

S=[.4943   .1494   .3563]

Section 9-3 Absorbing Markov Chains

Example 5: University Enrollment (page 461) Use a graphing calculator to find the limiting matrix.

Store the data in a matrix. Then raise the matrix to the 50th power.

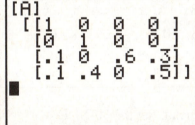 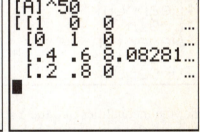

Scrolling to the right you see that the component of row 3 and column 3 is essentially a 0.

The matrix shows that the limiting matrix is $\begin{bmatrix} .4 & .6 \\ .2 & .8 \end{bmatrix}$.

From this you can determine that in the long run 60% of the first-year students will graduate and 20% of the second-year students will not graduate.

Chapter 10: Games and Decisions

For additional help on all aspects of the TI-83 or TI-84 graphing calculator, go to
http://www.prenhall.com/divisions/esm/app/graphing/ti83/

Section 10-1 Strictly Determined Games

There are no problems in this section that can be done with a graphing calculator.

Section 10-2 Mixed Strategy Games

There are no problems in this section that can be done with a graphing calculator.

Section 10-3 Linear Programming and 2 X 2 Games: Geometric Approach

Example 1: Solving 2 X 2 Matrix Games Using Geometric Methods (page 493).

(A) Minimize the equation $y = x_1 + x_2$.

Enter the inequality in part (A) using the application: Inequalz. Press APPS and scroll to the application.

You will need to rewrite the inequalities in terms of x and y in order to graph. **Refer to chapter 5, section 2, for complete instructions on using the inequality application.**

```
APPLICATIONS
 ↑GeoMastr
 ■Inequalz
 :LearnChk
 :LogIn
 :Nederlan
 :NoteFlio
 ↓OrganEsp
```

$$2x_1 + 5x_2 \geq 1 \implies 2x + 5y \geq 1 \implies y \geq \frac{1-2x}{5}$$

$$8x_1 + x_2 \geq 1 \implies 8x + y \geq 1 \implies y \geq 1 - 8x$$

$$x_1 \geq 0 \implies x \geq 0$$

$$x_2 \geq 0 \implies y \geq 0$$

Enter the inequalities into the Y= menu. Set the WINDOW.

```
X= Plot1  Plot2  Plot3
▼Y₁■(1-2X)/5
▼Y₂■1-8X
▼Y₃■0
\Y₄=
\Y₅=
\Y₆=
\Y₇=
```

```
■ Plot1  Plot2  Plot3
▼X₁■0■
\X₂=
\X₃=
\X₄=
\X₅=
\X₆=
```

```
WINDOW
 ShadeRes=3
 Xmin=0
 Xmax=1.1
 Xscl=.1
 Ymin=0
 Ymax=1
↓Yscl=.1
```

Graph the inequalities and then select intersection.

Use the TRACE key to find the corners.

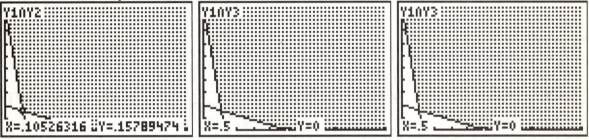

(B) Maximize the inequalities in this part as in part (A).

Section 10-4 Linear Programming and *m* X *n* Games: Simplex Method and the Dual Problem

There are no new calculator concepts in this section. Refer to section 6-2 on how to solve a problem using the Simplex method.

Chapter 11: Data Description and Probability Distributions

For additional help on all aspects of the TI-83 or TI-84 graphing calculator, go to http://www.prenhall.com/divisions/esm/app/graphing/ti83/

The TI-84 has many built-in statistical functions and graphing capabilities.

Section 11-1 Graphing Data

Example 2: Constructing Histograms with a Graphing Utility (page 510)

You can clear any data previously entered in a list by pressing STAT 4:ClrList and then enter the names of the lists. You can enter the list names by pressing 2nd STAT NAMES.

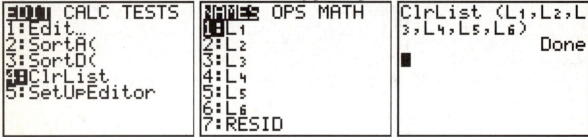

Now enter the data into list L1. Press STAT 1:Edit. Now turn on the STAT PLOTS by pressing 2nd Y= and press ENTER.

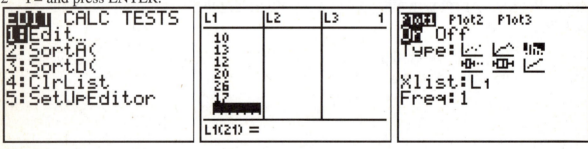

Set the window and turn on the grid by going to FORMAT. Press GRAPH to display the histogram.

To determine the frequency of each interval, use the TRACE key.

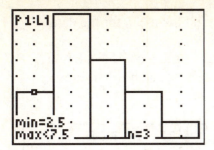

Exercise 11.1.9 (page 513) Railroad Freight

There are several types of graphs that can be produced in the STAT PLOT menu. The directions for this problem are to produce a broken-line graph. For the first year, 1940, set X=0, and then in increments of 10 in L1. The car loadings are then put in L2. In the STAT PLOT, select the line graph.

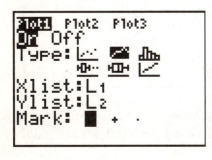

Section 11-2 Measures of Central Tendency

Example 1: Find the Mean (page 517). Find the mean for the sample measurements 3,5,1,8,6,5,4, and 6.

Enter the data into L1. Press STAT and go to the CALC menu. Select 1: 1-Var Stats, and then select L1. Press ENTER. The statistics for L1 are displayed. The average is shown as 4.75.

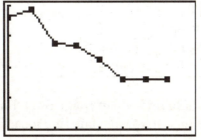

Example 2: Finding the Mean for Grouped Data (page 518). Find the mean for the data shown in the table.

Put the midpoint of each class interval in L1 and put the frequency in L2. Now select STAT, CALC, 1-Var Stats and then enter L1 as the list of data, and L2 as the frequency list. Press ENTER and read the average to be 579.

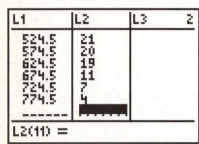 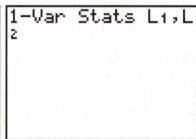 

Example 3: Finding the Median (page 519). Find the median salary by entering the salaries in L1.

This can be done on the home screen as shown. Scroll down to locate the median of the list.

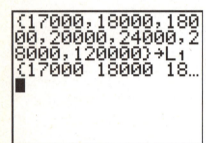  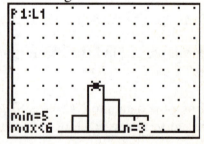

Example 5: Finding Mode, Median, and Mean (page 521). Enter the data in L1. You can find the mean and median by going to the catalog and entering mean(L1) and median(L1). To find the mode construct a histogram.

The mode is found in the histogram by noting the tallest block on the histogram.

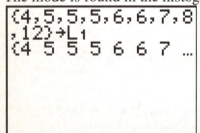 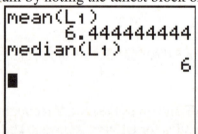

Section 11-3 Measures of Dispersion

Example 1: Finding the Standard Deviation (page 527). Find the standard deviation for the sample measurements 1, 3, 5, 4, 3.

Put the numbers in L1. Go the STAT menu and select 1: 1-Var Stats. The sample standard deviation of x is listed as Sx. We can see the value Sx=1.48. The population standard deviation, σx, is also shown if the sample is considered to be the entire population.

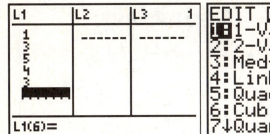 

Example 2: Finding the Standard Deviation for Grouped Data (page 528). (A) Find the standard deviation for the grouped sample data shown in the histogram.

There is no list of data given, so you will read the midpoint of each group and place that value in L1. Then you will put the frequency for each group in L2. You will calculate using 1-Var Stats.

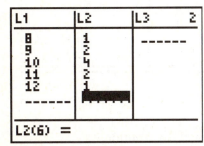 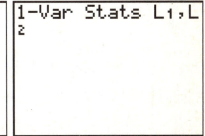

You can then take note of any of the statistics that you want. In this case, we are finding the standard deviation of 1.15.

```
1-Var Stats
 x̄=10
 Σx=100
 Σx²=1012
 Sx=1.154700538
 σx=1.095445115
↓n=10
```

Section 11-4 Bernoulli Trials and Binomial Distributions

Example 5: Constructing Tables and Histograms for Binomial Distributions (page 535).
Expand the problem to 100 repetitions and use the graphing calculator to generate the data.

Press MATH and scroll over to the PRB (probability) menu. Select 7: randBin (random binomial). The probability of rolling a number divisible by 3 is 1/3. The numbers are entered

(3,1/3,100) which means that the die is thrown three times, the probability of success each time is 1/3, and the experiment is to be repeated 100 times. The results are stored in L1. To graph the histogram, press 2nd Y= which is the STAT PLOT menu.

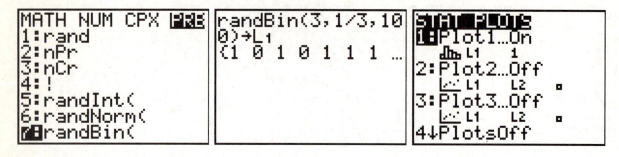

Turn Plot1 on, and select the histogram. Set the window. Create the histogram. Use TRACE to see the number of occurrences.

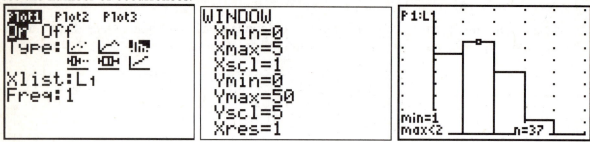

Section 11-5 Normal Distributions

A graphing calculator will generate random numbers from the normal distribution.

Example 1: Finding Probabilities for a Normal Distribution (page 543). Use a graphing calculator to generate 100 random numbers from the normal distribution. The entries are $(\mu, \sigma, \#trials)$. You can calculate the stats for the list using the STAT menu.

Now create a histogram to determine the empirical probability. Set the Xscl to 40 so that the histogram has the proper grouping. Use TRACE to find the values.

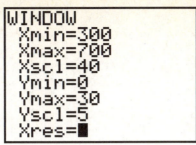
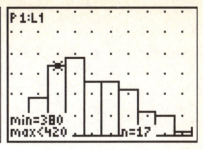

The frequencies found on the histogram are 17, 19, and 13. So the probability that a light bulb will last between 380 and 500 hours is found to be .49.

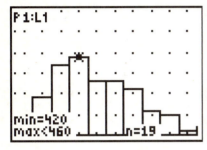
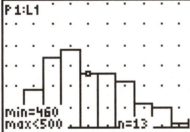
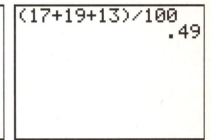

EXCEL® SPREADSHEET MANUAL

DALE R. BUSKE

St. Cloud State University

FINITE MATHEMATICS FOR BUSINESS, ECONOMICS, LIFE SCIENCES AND SOCIAL SCIENCES

TWELFTH EDITION

Prentice Hall
is an imprint of

CONTENTS

Chapter 1

Section 1-1 Linear Equations and Inequalities

Example 1: Solving a Linear Equation

While linear equations can be "solved" using Excel (see Example 2), for now we focus on checking solutions to algebraic equations. In this example, we verify that $x = -9$ is a solution to $8x - 3(x - 4) = 3(x - 4) + 6$.

Enter: Value **-9** in cell **A2.**
Enter: Formula **=8*A2-3*(A2-4)** in cell **B2.**

	A	B	C
1	x	LHS	RHS
2		-9	=8*A2-3*(A2-4)

Good Habit: The best way to enter such a formula in Excel is to type **=8*** in cell **B2.** Then, use the mouse to click on the contents in cell **A2.** Next, continue to type **-3*(** followed by clicking on the contents of cell **A2** once again followed by typing **-4)** in cell **B2.** Such a visual approach to formula entry leads to fewer typographical errors.

Enter: Formula **=3*(A2-4)+6** in cell **C2.**

	A	B	C	D
1	x	LHS	RHS	
2	-9		-33	=3*(A2-4)+6

Good Habit: Remember to press **ENTER** every time a formula is entered in a cell in Excel.

Since the resulting values in cells **B2** and **C2** match, $x = -9$ is a solution to $8x - 3(x - 4) = 3(x - 4) + 6$.

	A	B	C
1	x	LHS	RHS
2	-9	-33	-33

Example 2: Solving a Linear Equation

We solve $\dfrac{x+2}{2} - \dfrac{x}{3} = 5$ for x.

Enter: Any starting value (e.g. 0) in cell **A2.**
Enter: Formula **=(A2+2)/2 − A2/3** in cell **B2.**

	A	B	C
1	x	LHS	
2	0	=(A2+2)/2 - A2/3	

Good Habit: Remember to press **ENTER** every time a formula is entered in a cell in Excel.

Select: On the **Data** tab, select **Goal Seek** under the **What-If-Analysis** menu.

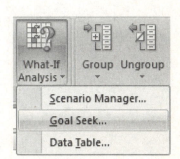

1

Enter: Set cell: **B2**
Enter: To value: **5**
Enter: By changing cell: **A2**
Click: **OK**

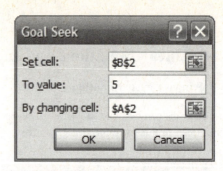

Good Habit: Use the mouse to actually click on cells **B2** and **A2** for "Set cell:" and "By changing cell:" rather than typing the actual cell locations.

Goal Seek Status window will appear.
Click: **OK**

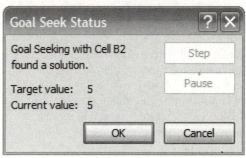

A "solution" will appear in cell **A2.** In general, when solving linear equations using **Goal Seek** in Excel the solution that appears will not be exact.

	A	B
1	x	LHS
2	24	5

Example 8: Purchase Price

Enter: Any value (e.g. **1**) for the unknown purchase price *x* in cell **A2.**
Enter: The value **57** in cell **B2.**
Enter: Formula **=0.052*A2** in cell **C2.**
Enter: Formula **=SUM(A2:C2)** in cell **D2.**

	A	B	C	D
1	x	Shipping	Tax	Total
2	1	57	=0.052*A2	

	A	B	C	D	E
1	x	Shipping	Tax	Total	
2	1	57	0.052	=SUM(A2:C2)	

Select: On the **Data** tab, select **Goal Seek** under the **What-If-Analysis** menu.

Enter: Set cell: **D2**
Enter: To value: **851.26**
Enter: By changing cell: **A2**
Click: **OK**

Good Habit: Use the mouse to actually click on cells **D2** and **A2** for "Set cell:" and "By changing cell:" rather than typing the actual cell locations.

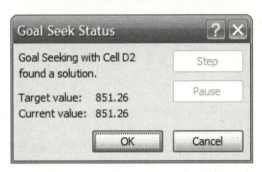

Goal Seek Status window will appear.
Click: **OK**

The solution (the unknown purchase price) will appear in cell **A2.**

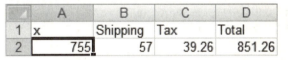

	A	B	C	D
1	x	Shipping	Tax	Total
2	755	57	39.26	851.26

Example 9: Break-Even Analysis

Enter: Any value (e.g. **1**) for the unknown number of DVD's *x* in cell **A2.**
Enter: Value **48000** in cell **B2.**
Enter: Formula **=12.40*A2** in cell **C2.**
Enter: Formula **=SUM(B2:C2)** in cell **D2.**
Good Habit: Use the mouse to select cells **B2** through **C2** for the sum rather than typing **B2:C2**. Click on **B2** and drag over to **C2**.
Enter: Formula **=17.40*A2** in cell **E2.**

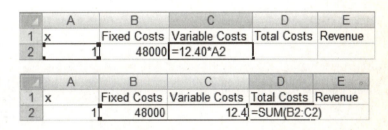

	A	B	C	D	E
1	x	Fixed Costs	Variable Costs	Total Costs	Revenue
2	1	48000	=12.40*A2		

	A	B	C	D	E
1	x	Fixed Costs	Variable Costs	Total Costs	Revenue
2	1	48000	12.4	=SUM(B2:C2)	

	A	B	C	D	E
1	x	Fixed Costs	Variable Costs	Total Costs	Revenue
2	1	48000	12.4	48012.4	=17.4*A2

Enter: Formula **=E2-D2** in cell **F2.**

	A	B	C	D	E	F
1	x	Fixed Costs	Variable Costs	Total Costs	Revenue	Revenue-Costs
2	1	48000	12.4	48012.4	17.4	=E2-D2

Select: On the **Data** tab, select **Goal Seek** under the **What-If-Analysis** menu.

Enter: Set cell: **F2**
Enter: To value: **0**
Enter: By changing cell: **A2**
Click: **OK**

Good Habit: Use the mouse to actually click on cells **F2** and **A2** for "Set cell:" and "By changing cell:" rather than typing the actual cell locations.

Goal Seek Status window will appear.
Click: **OK**

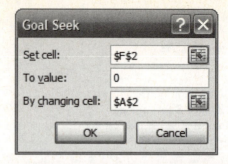

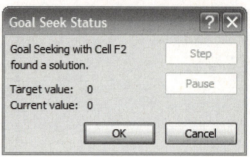

The solution (the unknown number of DVDs) will appear in cell **A2**.

	A	B	C	D	E	F
1	x	Fixed Costs	Variable Costs	Total Costs	Revenue	Revenue-Costs
2	9600	48000	119040	167040	167040	0

Example 10: Consumer Price Index

Enter: CPI data from chart.
Enter: Any value for the unknown salary *x* in cell **C3** (e.g. **50,000**).

	A	B	C
1		1960	2005
2	CPI	29.6	195.3
3	Salary	13000	50000

Enter: Formula **=C2/B2** in cell **E2**.

	A	B	C	D	E
1		1960	2005		Ratio
2	CPI	29.6	195.3		=C2/B2
3	Salary	13000	50000		

Copy: Content of cell **E2** to cell **E3**. The formula **=C3/B3** will then appear in cell **E3**.

	A	B	C	D	E
1		1960	2005		Ratio
2	CPI	29.6	195.3		6.597973
3	Salary	13000	50000		=C3/B3

Enter: Formula **=E3-E2** in cell **F3**. [This is the difference of the ratios which we will set to 0.]

	A	B	C	D	E	F
1		1960	2005		Ratio	Difference
2	CPI	29.6	195.3		6.597973	
3	Salary	13000	50000		3.846154	=E3-E2

Select: On the **Data** tab, select **Goal Seek** under the **What-If-Analysis** menu.

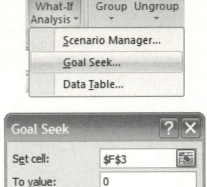

Enter: Set cell: **F3**
Enter: To value: **0**
Enter: By changing cell: **C3**
Click: **OK**

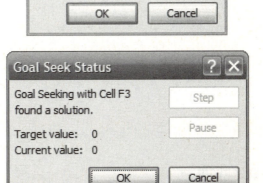

Good Habit: Use the mouse to actually click on cells **F2** and **A2** for "Set cell:" and "By changing cell:" rather than typing the actual cell locations.

Goal Seek Status window will appear.
Click: **OK**

The solution will appear in cell **C3**.

	A	B	C	D	E	F
1		1960	2005		Ratio	Difference
2	CPI	29.6	195.3		6.597973	
3	Salary	13000	85773.65		6.597973	0

Section 1-2 Graphs and Lines

Example 2: Using a Graphing Calculator

To generate *x*-values starting at -10 and increasing by 1,
Enter: The value **-10** in cell **A2**.
Enter: Formula **=A2+1** in cell **A3**.

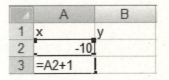

Good Habit: Rather than typing cell reference **A2** in the formula, use the mouse to click on cell **A2** to avoid typos.
Copy: Contents of cell **A3** to cells **A4:A22**.

Good Habit: Hold the left mouse button while over the lower right corner of cell **A3** (make sure you point the arrow exactly on the lower right corner of the cell until you see **+**), and drag the mouse down to cell **A22**.

Enter: Formula **=3/4*A2-3** in cell **B2.**

Copy: Contents of cell **B2** to cells **B3:B22.**

Good Habit: Hold the left mouse button while over the lower right corner of cell **B2**, and drag the mouse down to cell **B22**.

Select: Cells **A2:B22**

Select: On the **Insert** tab, select **Scatter** on the **Charts** menu. Then, select **Scatter with Smooth Lines and Markers**.

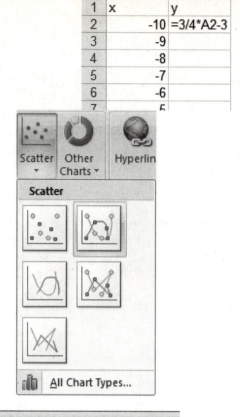

	A	B
1	x	y
2	-10	=3/4*A2-3
3	-9	
4	-8	
5	-7	
6	-6	
7	-5	

Select: **Design, Layout,** or **Format** tab under the **Chart Tools** menu. Use the submenus to format the output as desired.
For example,

Select: **Layout** tab and click on **Gridlines** on the **Axes** submenu. Then, select **Primary Horizontal Gridlines** and **None** to eliminate any horizontal gridlines.

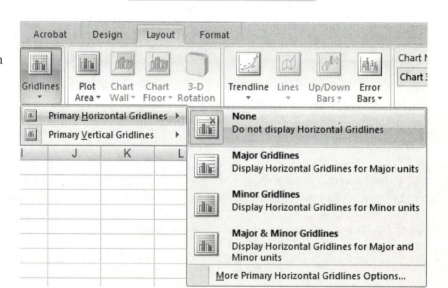

For example,

Select: **Layout** tab and click on **Axes** on the **Axes** submenu. Then, select **Primary Horizontal Axis** and **More Primary Horizontal Axis Options**.

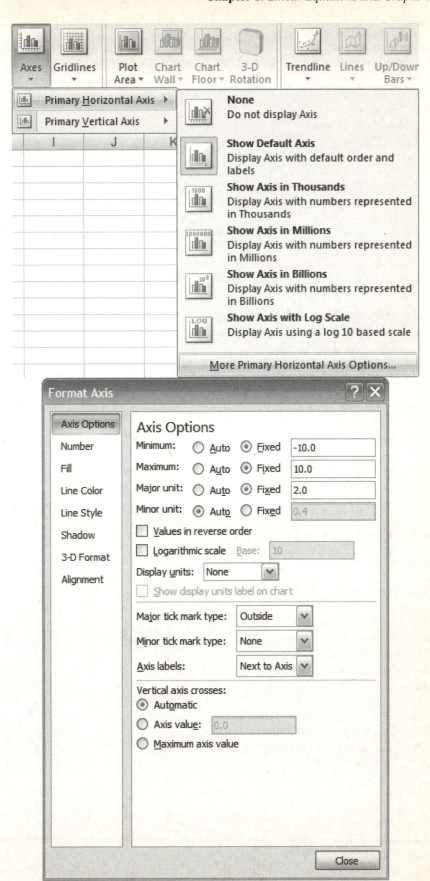

The **Format Axis** window will pop up. Select **Minimum** to be **Fixed** at **-10.0**. Select **Maximum** to be **Fixed** at **10.0**. Select **Major unit** to be **Fixed** at **2.0**.

Using the **Chart Tools**, output can be formatted to look however you wish. One possibility is shown at the right.

Note: Right clicking on the graph itself also allows for editing of the output.

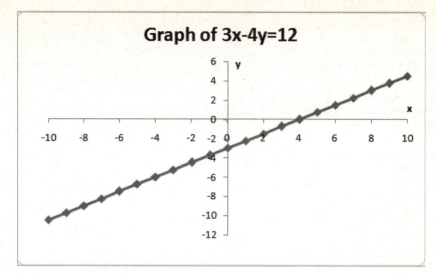

Example 4: Finding Slopes

Enter: The value **-3** in cell **A2**.
Enter: The value **3** in cell **A3**.
Enter: The value **-2** in cell **B2**.
Enter: The value **4** in cell **B3**.
Enter: Formula **=(B3-B2)/(A3-A2)** in cell **D2**.

	A	B	C	D	E
1	x	y		Slope	
2	-3	-2		=(B3-B2)/(A3-A2)	
3	3	4			

To compute the slope of a new line, simply change the values of the ordered pairs in cells **A2**, **A3**, **B2**, and **B3**.

Enter: The value **-1** in cell **A2**.
Enter: The value **2** in cell **A3**.
Enter: The value **3** in cell **B2**.
Enter: The value **-3** in cell **B3**.

	A	B	C	D
1	x	y		Slope
2	-1	3		-2
3	2	-3		

Example 8: Supply and Demand

Input the data from the text into a tabular format on the worksheet.
To find the slope of the price-supply equation,
Enter: Formula **=(C2-C3)/(A2-A3)** in cell **A6**.
To find the slope of the demand equation,
Enter: Formula **=(C2-C3)/(B2-B3)** in cell **B6**.

	A	B	C
1	Supply (thousand boxes)	Demand (thousand boxes)	Price ($/box)
2	320	200	$9.00
3	270	300	$8.50
4			
5	Slope (Supply)	Slope (Demand)	
6	0.01	=(C2-C3)/(B2-B3)	

Note: To format the content in cells **C2** and **C3** as **Currency**, highlight these cells and select the Currency option on the **Number** menu under the **Home** tab.

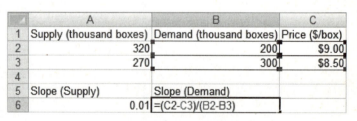

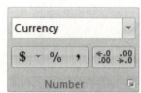

Enter: Any quantity (e.g. **100** boxes) in cell **A9**.

The information for the slope-intercept form of the price-supply equation is readily available.

Enter: Formula **=A6*(A9-A2)+C2** in cell **B9**.

Similarly, enter information for the slope-intercept form of the price-demand equation.

Enter: Formula **=B6*(A9-B2)+C2** in cell **C9**.

Enter: Formula **=B9-C9** in cell **D9**.

Our goal will be to determine for what quantity (*x* representing thousands of boxes) this value (the difference between the supply and demand) is equal to 0.

	A	B	C	D
1	Supply (thousand boxes)	Demand (thousand boxes)	Price ($/box)	
2	320	200	$9.00	
3	270	300	$8.50	
4				
5	Slope (Supply)	Slope (Demand)		
6	0.01	-0.005		
7				
8	x (Quantity)	Supply	Demand	Difference
9	280	$8.60	=B6*(A9-B2)+C2	

	A	B	C	D
1	Supply (thousand boxes)	Demand (thousand boxes)	Price ($/box)	
2	320	200	$9.00	
3	270	300	$8.50	
4				
5	Slope (Supply)	Slope (Demand)		
6	0.01	-0.005		
7				
8	x (Quantity)	Supply	Demand	Difference
9	280	$8.60	$8.60	=B9-C9

Select: On the **Data** tab, select **Goal Seek** under the **What-If-Analysis** menu.

Enter: Set cell: **D9**
Enter: To value: **0**
Enter: By changing cell: **A9**
Click: **OK**

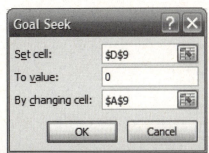

Goal Seek Status window will appear.
Click: **OK**

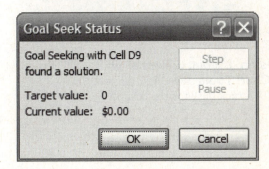

The equilibrium quantity will appear in cell **A9**. The equilibrium price will appear in cells **B9** and **C9**.

	A	B	C	D
1	Supply (thousand boxes)	Demand (thousand boxes)	Price ($/box)	
2	320	200	$9.00	
3	270	300	$8.50	
4				
5	Slope (Supply)	Slope (Demand)		
6	0.01	-0.005		
7				
8	x (Quantity)	Supply	Demand	Difference
9	280	$8.60	$8.60	$0.00

Section 1-3 Linear Regression

Example 3: Diamond Prices

Input the data from the text into tabular form on the worksheet.

	A	B
1	Weight	Price
2	0.5	2790
3	0.6	3191
4	0.7	3694
5	0.8	4154
6	0.9	5018
7	1	5898

Select: Cells **A2:B7**

Select: On the **Insert** tab, select **Scatter** on the **Charts** menu. Then, select **Scatter with only Markers**.

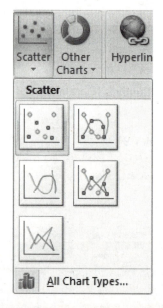

Select: **Design, Layout,** or **Format** tab under the **Chart Tools** menu. Use the submenus to format the output as desired.

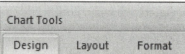

Sample output is shown to the right.

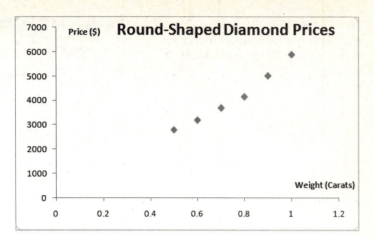

Select: **Layout** tab under the **Chart Tools** menu.

Select: **More Trendline Options** under the **Trendline** submenu.

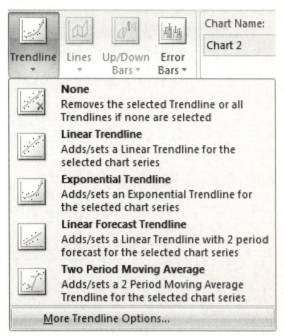

Select: **Linear** Trend/Regression Type
Select: **Display Equation on chart**
Select: **Display R-squared value on chart**
Click: **Close**

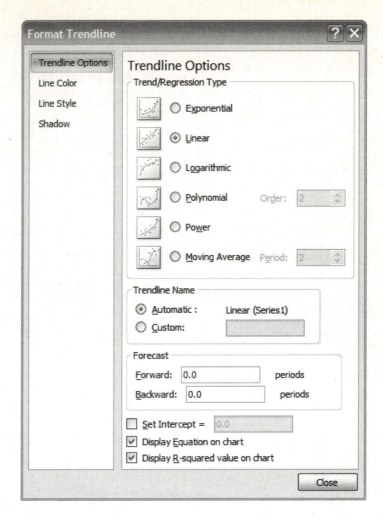

The regression line (a linear model for the data) will appear on the chart.

Here, *x* represents the value of the independent variable (Weight measured in Carats) and *y* represents the value of the dependent variable (Price in $).

Note: It is easy to edit the equation for the regression line so that it is given in terms of *p* (price) and *c* (carats).

Also, the slope and *y*-intercept have been rounded to the nearest tenth by a right mouse click and the use of **Format Trendline Label**.

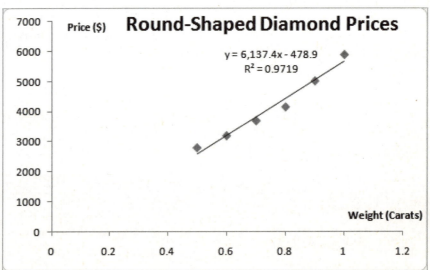

To estimate the value of 0.85 and 1.2 carat diamonds,
Enter: Value **0.85** in cell **A9**
Enter: Formula **=6137.4*A9-478.9** in cell **B9**

Then, enter the value **1.2** in cell **A9** to estimate the price of a 1.2 carat diamond.

Note: These results will vary from those in the text since that model has rounded the slope and *y*-intercept to the nearest ten (not tenth).

	A	B	C
1	Weight	Price	
2	0.5	2790	
3	0.6	3191	
4	0.7	3694	
5	0.8	4154	
6	0.9	5018	
7	1	5898	
8			
9	0.85	=6137.4*A9-478.9	

To estimate the weight of a $4,000 diamond,
Enter: *Any* guess (**e.g. 0.75**) to the value of such a diamond in cell **A9**

	A	B
1	Weight	Price
2	0.5	2790
3	0.6	3191
4	0.7	3694
5	0.8	4154
6	0.9	5018
7	1	5898
8		
9	0.75	4124.15

Select: On the **Data** tab, select **Goal Seek** under the **What-If-Analysis** menu.

Enter: Set cell: **B9**
Enter: To value: **4000**
Enter: By changing cell: **A9**
Click: **OK**

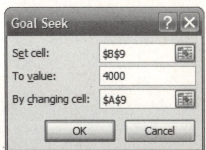

Goal Seek Status window will appear.
Click: **OK**

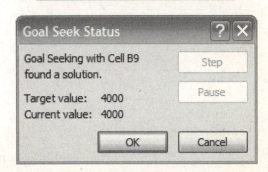

The estimated weight of a $4000 diamond (in Carats) will appear in cell **A9.**

	A	B
1	Weight	Price
2	0.5	2790
3	0.6	3191
4	0.7	3694
5	0.8	4154
6	0.9	5018
7	1	5898
8		
9	0.729772	4000

Chapter 2

Section 2-1 Functions

Example 1: Point-by-Point Plotting

To generate a list of *x*-values for plotting,
Enter: **-4** in cell **A2.**
Enter: Formula **=A2+1** in cell **A3.**

Good Habit: Rather than typing the cell reference **A2** in the formula, use the mouse to click on cell **A2** to avoid typos.

	A	B
1	x	y
2	-4	
3	=A2+1	

Copy: Contents of cell **A3** to cells **A4:A10.**

Good Habit: Hold the left mouse button while over the lower right corner of cell **A3**, and drag the mouse down to cell **A10.**

	A	B
1	x	y
2	-4	
3	-3	
4	-2	
5	-1	
6	0	
7	1	
8	2	
9	3	
10	4	

To generate a list of *y*-values for the function $y = 9 - x^2$,
Enter: Formula **=9-A2^2** in cell **B2.**

	A	B
1	x	y
2	-4	=9-A2^2
3	-3	
4	-2	

Copy: Contents of cell **B2** to cells **B3:B10.**

Good Habit: Hold the left mouse button while over the lower right corner of cell **B2**, and drag the mouse down to cell **B10.**

	A	B
1	x	y
2	-4	-7
3	-3	0
4	-2	5
5	-1	8
6	0	9
7	1	8
8	2	5
9	3	0
10	4	-7

Select: Cells **A2:B10**

Select: On the **Insert** tab, select **Scatter** on the **Charts** menu. Then, select **Scatter with Smooth Lines and Markers**.

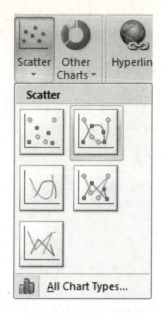

Select: **Design, Layout,** or **Format** tab under the **Chart Tools** menu. Use the submenus to format the output as desired.
For example,

Select: **Layout** tab and click on **Gridlines** on the **Axes** submenu. Then, select **Primary Horizontal Gridlines** and **Minor Gridlines** to add horizontal gridlines.

Select: **Layout** tab and click on **Gridlines** on the **Axes** submenu. Then, select **Primary Vertical Gridlines** and **Minor Gridlines** to add vertical gridlines.

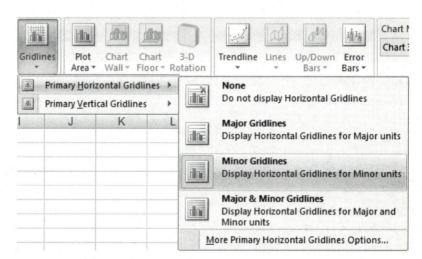

For example,

Select: **Layout** tab and click on **Axes** on
the **Axes** submenu. Then, select **Primary
Horizontal Axis** and **More Primary
Horizontal Axis Options...** to format
the horizontal axis.

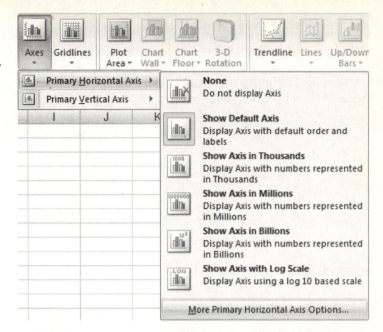

In the **Format Axis** window,
Select: **Line Style** on the left menu.
Select: **Width: 2 pt**
Select: **Begin type** arrow and **End type**
arrow under **Arrow settings**.
Click: **Close**

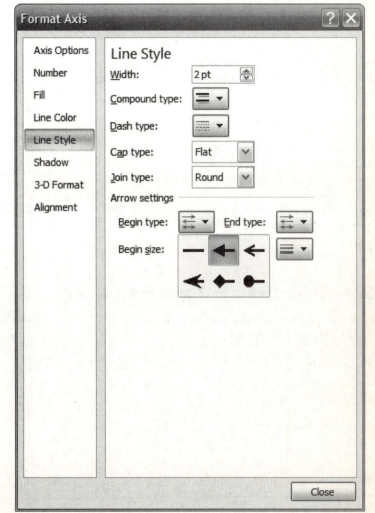

Using the **Chart Tools**, output can be formatted to look however you wish. One possibility is shown at the right.

Note: Right clicking on the graph itself also allows for editing of the output.

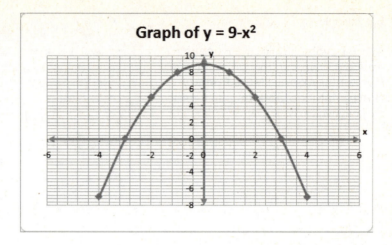

Graph of y = 9-x²

Example 4: Function Evaluation

To evaluate *f*(6) for the function $f(x) = \dfrac{12}{x-2}$,

Enter: Value **6** in cell **A2**.
Enter: Formula **=12/(A2-2)** in cell **B2**.
Good Habit: Rather than typing the cell reference **A2** in the formula, use the mouse to click on cell **A2** to avoid typos.
Note: The function value in cell **B2** will automatically update when new values (such as **4** shown at the right) are entered into cell **A2**.

	A	B
1	x	h(x)
2	6	=12/(A2-2)

	A	B
1	x	h(x)
2	4	6

To evaluate *h*(-2) for the function $h(x) = \sqrt{x-1}$,

Enter: Value **-2** in cell **A2**.
Enter: Formula **=SQRT(A2-1)** in cell **B2**.

Since **-2** is not in the domain of the function, an error message occurs in cell **B2**.

	A	B	C
1	x	h(x)	
2	-2	=SQRT(A2-1)	

	A	B
1	x	h(x)
2	-2	#NUM!

Example 7: Price-Demand and Revenue Modeling

Input the price-demand data from the text into a table on the worksheet.

	A	B
1	x (Millions)	p ($)
2	2	87
3	5	68
4	8	53
5	12	37

Select: Cells **A2:B5**

On the **Insert** tab, select **Scatter** on the **Charts** menu. Then, select **Scatter with only Markers**.

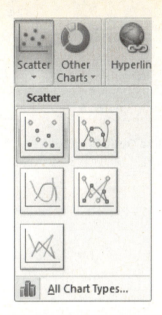

Select: **Design, Layout,** or **Format** tab under the **Chart Tools** menu. Use the submenus to format the output as desired.

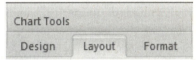

Sample output is shown to the right.

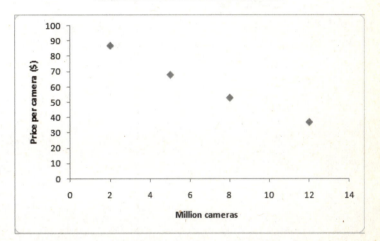

Select: **Layout** tab under the **Chart Tools** menu.

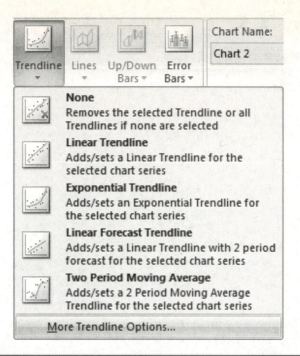

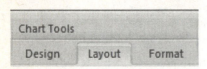

Select: **More Trendline Options** under the **Trendline** submenu.

Select: **Linear** Trend/Regression Type
Select: **Display Equation on chart**
Click: **Close**

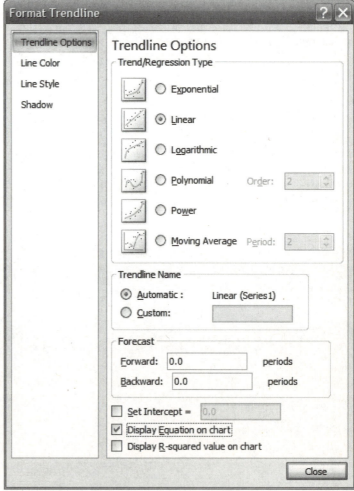

A regression line (a linear model for the data) will appear on the chart.

Here, *x* represents the value of the independent variable (Million cameras) and y represents the value of the dependent variable (Price per camera in $).

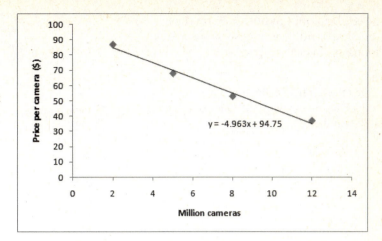

Input the *x*-values from the text into a table on the spreadsheet. We calculate the revenue $R(x) = x(-5x + 94.8)$ at these values.

Enter: Formula **=A2*(-5*A2+94.8)** in cell **B2**
Copy: The contents of cell **B2** to cells **B3:B7**

Good Habit: Hold the left mouse button while over the lower right corner of cell **B2**, and drag the mouse down to cell **B7**.

Select: Cells **A2:B7**.

	A	B	C
1	x (Millions)	R(x) (Million $)	
2	1	=A2*(-5*A2+94.8)	
3	3		
4	6		
5	9		
6	12		
7	15		

	A	B
1	x (Millions)	R(x) (Million $)
2	1	89.8
3	3	239.4
4	6	388.8
5	9	448.2
6	12	417.6
7	15	297

On the **Insert** tab, select **Scatter** on the **Charts** menu. Then, select **Scatter with Smooth Lines and Markers**.

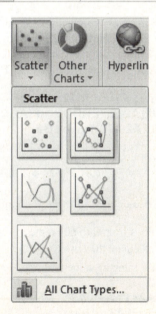

Select: **Design, Layout,** or **Format** tab under the **Chart Tools** menu. Use the submenus to format the output as desired.

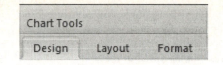

Using the **Chart Tools**, output can be formatted to look however you wish. One possibility is shown at the right.

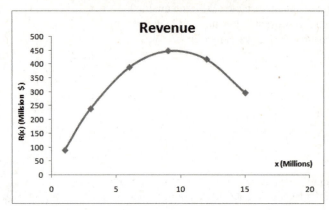

Section 2-2 Elementary Functions: Graphs and Transformations

Example 1: Evaluating Basic Elementary Functions

Mathematical Function	Excel Function
x^2	=x^2
x^3	=x^3
$\sqrt{x}$	=SQRT(x)
$\sqrt[3]{x}$	=x^(1/3)
$|x|$	=ABS(x)

Example 6: Natural Gas Rates

Piecewise defined functions can be implemented in Excel using the built-in IF function. The syntax of this function works as follows: IF(logical_test,value_if_true,value_if_false). In the following, Excel first checks whether the contents of cell **A2** has a value less than or equal to 0. If it does, then it outputs an error. Next, it checks if the contents have a value less than or equal to 5. Then, it checks to see if the contents are less than or equal to 40.

Enter: Any value (e.g. **7**) in cell **A2.**
Enter: Formula **=IF(A2<0,"ERROR",IF(A2<=5,0.7866*A2, IF(A2<=40, 3.933+0.4601*(A2-5), 20.0365+0.2508*(A2-40))))** in cell **B2.**

	A	B	C	D	E	F	G	H	I	J	K
1	x	C(x)									
2	7	=IF(A2<0,"ERROR",IF(A2<=5,0.7866*A2, IF(A2<=40, 3.933+0.4601*(A2-5), 20.0365+0.2508*(A2-40))))									

Graphing a piecewise-defined function is no different than graphing any other function in Excel.

Input various *x*-values from 0 to 60 on the spreadsheet. We calculate the cost $C(x)$
at these (in this case integer) values.
Enter: **0** in cell **A2.**
Enter: Formula **=A2+1** in cell **A3.**
Copy: Contents of cell **A3** to cells **A4:A62.**

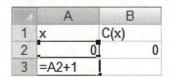

Copy: Contents of cell **B2** to cells
B3:B62.

Good Habit: Hold the left mouse button
while over the lower right corner of cell
B2, and drag the mouse down to cell **B62**.

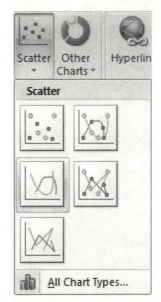

Select: Cells **A2:B62**.

Select: On the **Insert** tab, select **Scatter**
on the **Charts** menu. Then, select
Scatter with Smooth Lines.

Select: **Design, Layout,** or **Format** tab
under the **Chart Tools** menu. Use the
submenus to format the output as desired.

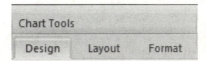

For example,

Select: **Layout** tab and click on **Axes** on the **Axes** submenu. Then, select **Primary Vertical Axis** and **More Primary Vertical Axis Options...** to format the vertical axis.

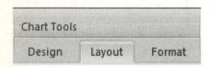

In the **Format Axis** window,
Select: **Number** on the left menu.
Select: **Category: Currency**
Enter: **Decimal places: 0**
Click: **Close**

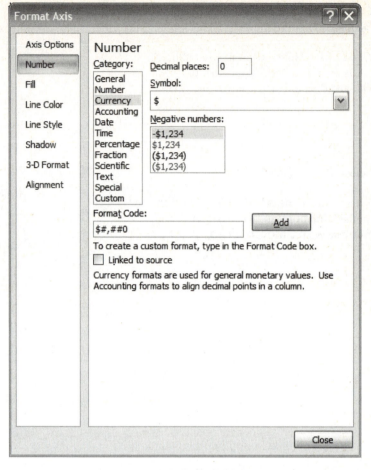

Using the **Chart Tools**, output can be formatted to look however you wish. One possibility is shown at the right.

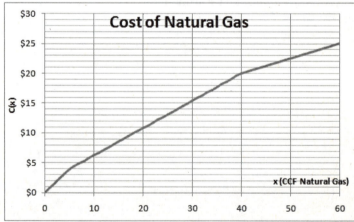

Section 2-3 Quadratic Functions

Example 1: Intercepts, Equations, and Inequalities

To find the *y*-intercept for the graph of $f(x) = -x^2 + 5x + 3$,
Enter: **0** in cell **A2**.
Enter: Formula **=(-1)*A2^2+5*A2+3** in cell **B2**.

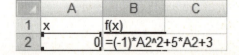

To find the *x*-intercepts using the quadratic formula,
Enter: Values of **a = -1, b = 5,** and **c = 3** in cells **A2, B2,** and **C2**
Enter: Formula **=(-B2+SQRT(B2^2-4*A2*C2))/(2*A2)** in cell **A5.**

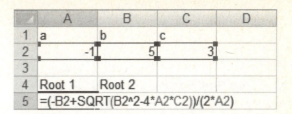

Enter: Formula **=(-B2-SQRT(B2^2-4*A2*C2))/(2*A2)** in cell **B5.**

The equation $-x^2 + 5x + 3 = 0$ could also be approximately solved using the **Goal Seek** function in Excel. We illustrate a similar example by estimating solutions to $-x^2 + 5x + 3 = 4$.

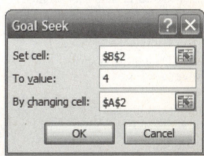

Enter: Any value (e.g. **0**) in cell **A2**.
Enter: Formula **=(-1)*A2^2+5*A2+3** in cell **B2**.
Select: On the **Data** tab, select **Goal Seek** under the **What-If-Analysis** menu.

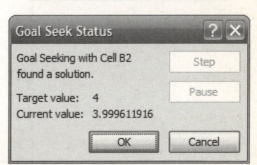

Enter: Set cell: **B2**
Enter: To value: **4**
Enter: By changing cell: **A2**
Click: **OK**

Good Habit: Use the mouse to actually click on cells **B2** and **A2** for "Set cell:" and "By changing cell:" rather than typing the actual cell locations.

Goal Seek Status window will appear.
Click: **OK**

Note: The Current value is only close to the Target value.

An approximate solution to $-x^2 + 5x + 3 = 4$ appears in cell **A2**.

	A	B
1	x	f(x)
2	0.208627	3.999612

Notice that **Goal Seek** only approximates one of the two solutions to $-x^2 + 5x + 3 = 4$. In fact, the Excel **Goal Seek** function is very sensitive to the initial value from cell **A2**.

Starting the Goal Seek with a value in cell A2 closer to the other solution, such as **6**, provides an approximate to the other solution.

Since we know there are only two solutions to this quadratic, we are reasonably comfortable with the two approximations $x \approx 0.21$ and $x \approx 4.79$.

	A	B
1	x	f(x)
2	4.791288	3.999998

Example 2: Analyzing a Quadratic Function

The **Solver** Add-in for Excel can help analyze quadratic and other functions. *If this Add-in has not been installed before*, you will need to install it.

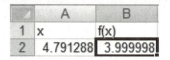

Select: **Office Button** (shown at right)
Select: **Excel Options** button at bottom

In the **Excel Options** window,

Select: **Add-Ins** on the left menu
Select: **Solver Add-in**
Click: **Go...**

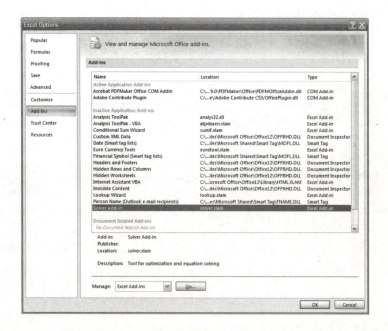

In the **Add-Ins** window,

Select: **Solver Add-in**
Click: **OK**

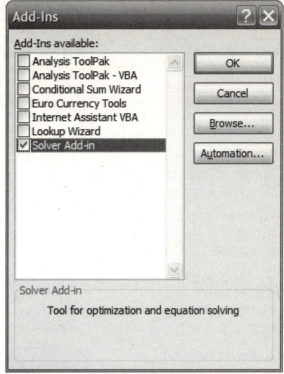

In the **Microsoft Office Excel** window,

Select: **Yes**

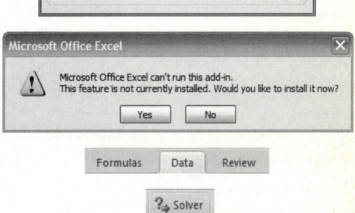

The **Solver** tool will appear under the
Data tab on the menu.

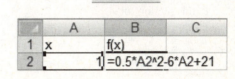

To find the vertex for the parabola
determined by the function
$f(x) = 0.5x^2 - 6x + 21$,

Enter: Any value (e.g. **1**) in cell **A2.**
Enter: Formula = **0.5*A2^2-6*A2+21** in
cell **B2.**

	A	B	C
1	x	f(x)	
2		1	=0.5*A2^2-6*A2+21

Select: **Data** tab

Click: **Solver**

Finding the vertex of a parabola is equivalent to finding its minimum if the parabola opens upward.

Enter: **Set Target Cell: B2**
Select: **Equal To: Min**
Enter: **By Changing Cells: A2**
Click: **Solve**

Select: **Keep Solver Solution**
Click: **OK**

Since the parabola opens upward, the coordinates for the vertex (the minimum found) will appear in cells **A2** and **B2**.

	A	B
1	x	f(x)
2	6	3

Example 3: Maximum Revenue

If the **Solver** tool is not present under the **Data** tab, see Example 2 in this section for instructions on installing.

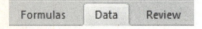

To maximize the revenue function
$R(x) = x(94.8 - 5x)$,

Enter: Any value (e.g. **3**) in cell **A2.**
Enter: Formula **=A2*(94.8-5*A2)** in cell **B2.**

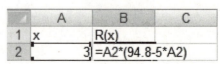

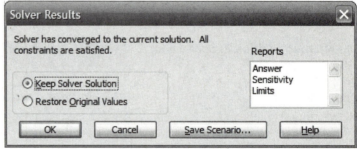

Select: **Data** tab

Click: **Solver**

In the **Solver Parameters** window,
Enter: **Set Target Cell: B2**
Select: **Equal To: Max**
Enter: **By Changing Cells: A2**
Click: **Subject to the Constraints: Add**

In the **Add Constraint** window,
Enter: **Cell Reference: A2**
Select: **>=**
Enter: **Constraint: 1**
Click: **Add**

In the **Add Constraint** window,
Enter: **Cell Reference: A2**
Select: **<=**
Enter: **Constraint: 15**
Click: **OK**

In the **Solver Parameters** window,
Click: **Solve**

In the **Solver Results** window,
Select: **Keep Solver Solution**
Click: **OK**

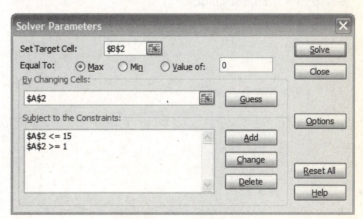

The coordinates of the maximum will appear in cells **A2** and **B2**. The maximum is the same as the vertex since the parabola $R(x) = x(94.8 - 5x)$ opens downward.

	A	B
1	x	R(x)
2	9.48	449.352

Example 4: Break-Even Analysis

To approximate the break-even points
when cost $C(x) = 156 + 19.7x$ and revenue

$R(x) = x(94.8 - 5x)$, when both have

domain $[1,15]$,
Enter: The value **1** in cell **A2**
Enter: Formula **=A2+1** in cell **A3**
Copy: Contents of cell **A3** to cells
A4:A16

	A	B	C
1	x	R(x)	C(x)
2	1		
3	=A2+1		

Enter: Formula **=A2*(94.8-5*A2)** in cell
B2.
Copy: Contents of cell **B2** to cells
B3:B16

Enter: Formula **=156+19.7*A2** in cell **C2.**
Copy: Contents of cell **C2** to cells
C3:C16

Good Habit: Hold the left mouse button
while over the lower right corner of cell
B2, and drag the mouse down to cell **B16**.
Do the same on column C.

	A	B	C	D
1	x	R(x)	C(x)	
2	1	89.8	=156+19.7*A2	
3	2	169.6		
4	3	239.4		
5	4	299.2		
6	5	349		
7	6	388.8		
8	7	418.6		
9	8	438.4		
10	9	448.2		
11	10	448		
12	11	437.8		
13	12	417.6		
14	13	387.4		
15	14	347.2		
16	15	297		

Select: Cells **A2:C16**.

Select: On the **Insert** tab, select **Scatter**
on the **Charts** menu. Then, select
Scatter with Smooth Lines.

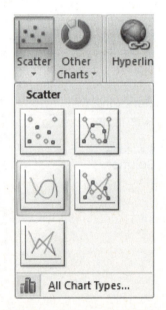

Select: **Design, Layout,** or **Format** tab
under the **Chart Tools** menu. Use the
submenus to format the output as desired.

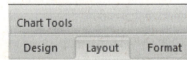

For example,

Select: **Layout** tab and click on **Legend** on the **Labels** submenu. Then, select **Overlay Legend at Right**.

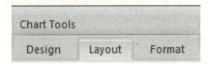

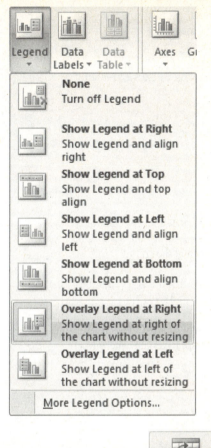

For example,

Select: **Design** tab and click on **Select Data** on the **Data** submenu.

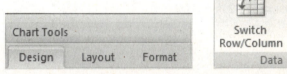

In **Select Data Source** window,
Select: **Series1**
Click: **Edit**

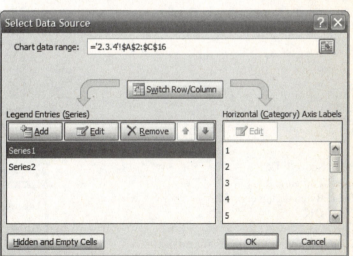

In the **Edit Series** window,
Enter: **Series name: R(x)**
Click: **OK**

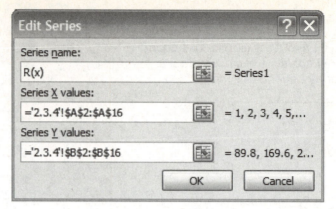

Repeat this process on **Series2** to label
C(x).

Using the **Chart Tools**, output can be
formatted to look however you wish. One
possibility is shown at the right.

The graph suggests that break-even points
are close to $x = 2.5$ and $x = 12.5$.

In what follows, we use the **Goal Seek**
Excel function to approximate where $R(x)$
$- C(x) = 0$.

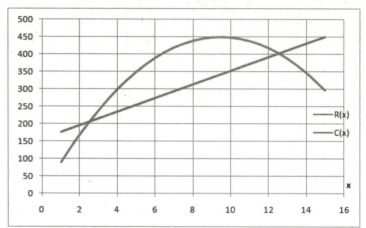

Enter: Formula **=B2-C2** in cell **D2**

	A	B	C	D
1	x	R(x)	C(x)	
2	1	89.8	175.7	=B2-C2
3	2	169.6	195.4	

Select: On the **Data** tab, select **Goal
Seek** under the **What-If-Analysis** menu.

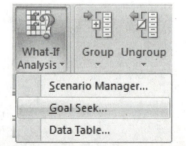

Enter: Set cell: **D2**
Enter: To value: **0**
Enter: By changing cell: **A2**
Click: **OK**

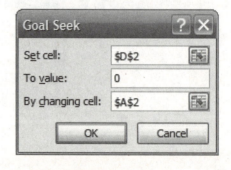

Good Habit: Use the mouse to actually
click on cells **D2** and **A2** for "Set cell:"
and "By changing cell:" rather than typing
the actual cell locations.

In the Goal Seek Status window,
Click: **OK**

An estimate for one of the two break-even points (the one closest to $x = 2$) will appear in cell **A2**.

	A	B	C	D
1	x	R(x)	C(x)	
2	2.49003	205.0536	205.0536	-1.1E-06

To estimate the other break-even point, Enter: A value close to the other break-even point in cell **A2**. (e.g. **12** or **13** found from the graph above)
Select: On the **Data** tab, select **Goal Seek** under the **What-If-Analysis** menu.
Enter: Set cell: **D2**
Enter: To value: **0**
Enter: By changing cell: **A2**
Click: **OK**

	A	B	C	D
1	x	R(x)	C(x)	
2	12.52997	402.8405	402.8404	0.000121

Example 5: Outboard Motors

Input the data from the text into a table on the worksheet.

	A	B	C
1	RPM	MPH	MPG
2	2500	10.3	4.1
3	3000	18.3	5.6
4	3500	24.6	6.6
5	4000	29.1	6.4
6	4500	33	6.1
7	5000	36	5.4
8	5400	38.9	4.9

Select: Cells **B2:C8**

Select: On the **Insert** tab, select **Scatter** on the **Charts** menu. Then, select **Scatter with only Markers**.

Select: **Design, Layout,** or **Format** tab under the **Chart Tools** menu. Use the submenus to format the output as desired.

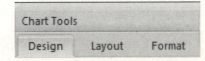

Using the **Chart Tools**, output can be formatted to look however you wish. One possibility is shown at the right.

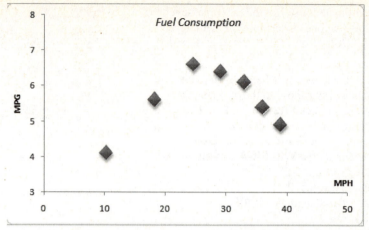

Select: **Layout** tab under the **Chart Tools** menu.

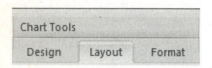

Select: **More Trendline Options** under the **Trendline** submenu.

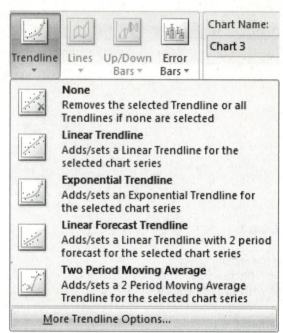

In the **Format Trendline** window,
Select: **Polynomial** Trend/Regression
Type
Select: **Order: 2** (for quadratic)
Select: **Display Equation on chart**
Click: **Close**

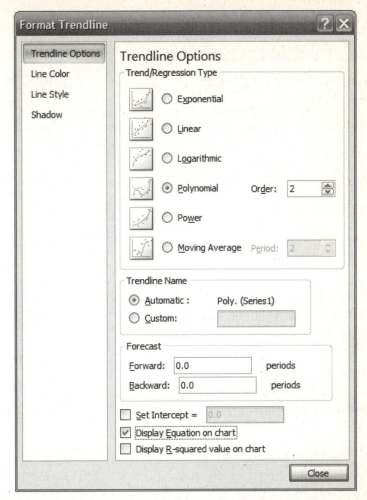

A regression line (a quadratic model for the data) will appear on the chart.

Here, *x* represents the value of the independent variable (MPH) and y represents the value of the dependent variable (MPG).

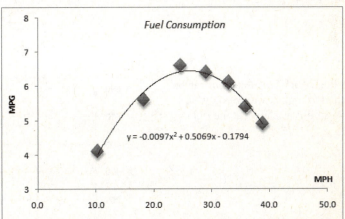

Fuel Consumption

$y = -0.0097x^2 + 0.5069x - 0.1794$

Section 2-4 Polynomial and Rational Functions

Example 1: Estimating the Weight of a Fish

Copy the data from the text into the worksheet.

	A	B
1	Length (in.)	Weight (oz.)
2	10	5
3	14	12
4	18	26
5	22	56
6	26	96
7	30	152
8	34	226
9	38	326
10	44	536

Select: Cells **A2:B10**

Select: On the **Insert** tab, select **Scatter** on the **Charts** menu. Then, select **Scatter with only Markers**.

Select: **Design, Layout,** or **Format** tab under the **Chart Tools** menu. Use the submenus to format the output as desired.

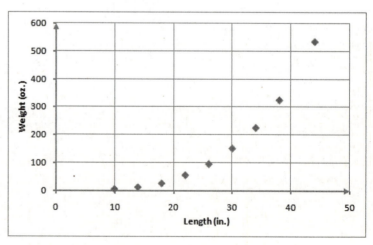

Using the **Chart Tools**, output can be formatted to look however you wish. One possibility is shown at the right.

Select: **Layout** tab under the **Chart Tools** menu.

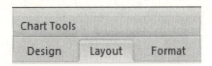

Select: **More Trendline Options** under the **Trendline** submenu.

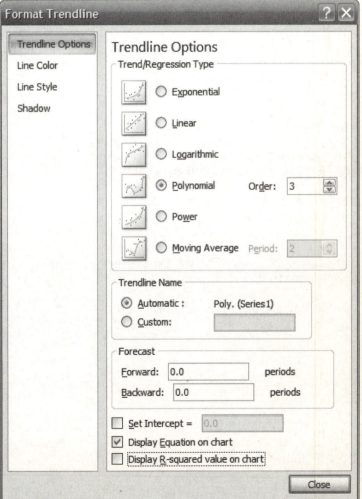

In the **Format Trendline** window, Select: **Polynomial** Trend/Regression Type
Select: **Order: 3** (for a cubic polynomial)
Select: **Display Equation on chart**
Click: **Close**

Note: In this case, since the weight of a 0 in. fish would obviously be 0 oz., the **Set Intercept = 0.0** box could also logically be selected to produce a good cubic model.

A regression line (a cubic model for the data) will appear on the chart.

Here, *x* represents the value of the independent variable (length of fish in inches) and *y* represents the value of the dependent variable (weight of fish in ounces).

Note that two different cubic models can be determined (depending on whether the intercept is set at 0.0 or not). If the intercept is not set at 0.0, the model is as shown at the right. However, if the intercept were set at 0.0, the cubic model produced would be

$y = 0.0095x^3 - 0.1179x^2 + 0.9268x$.

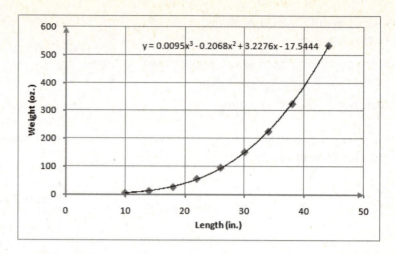

$y = 0.0095x^3 - 0.2068x^2 + 3.2276x - 17.5444$

Example 2: Graphing Rational Functions

We first enter *x*-values starting at -7 and increasing by, say, 0.5 until 7.

Enter: The value **-7** in cell **A2**.
Enter: Formula **=A2+0.5** in cell **A3**.
Copy: Contents of cell **A3** to cells **A4:A30**.

Alternative: Enter values -7 and -6.5 in cells **A2** and **A3** respectively. Highlight both cells and hold down the left mouse button while over the lower right corner of cell **A3**, Then, drag the mouse down to cell **A30**.

Enter: Formula **=3*A2/(A2^2-4)** in cell **B2**.
Copy: Contents of cell **B2** to cells **B3:B30**.

Good Habit: Hold down the left mouse button while over the lower right corner of cell **B2**, and drag the mouse down to cell **B30**.

Note: Cells **B12** and **B20** return division by zero errors.

	A	B	C
1	x	f(x)	
2		-7	=3*A2/(A2^2-4)
3		-6.5	
4		-6	
5		-5.5	
6		-5	

	A	B
1	x	f(x)
11	-2.5	-3.33333
12	-2	#DIV/0!
13	-1.5	2.571429

Select: Cells **A2:B30**

Select: On the **Insert** tab, select **Scatter** on the **Charts** menu. Then, select **Scatter with Smooth Lines and Markers**.

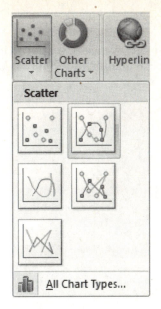

Select: **Design, Layout,** or **Format** tab under the **Chart Tools** menu. Use the submenus to format the output as desired.

User Beware: Notice that the graph has key errors. It does not illustrate (well) the vertical asymptotes at $x = -2$ and $x = 2$. In fact, Excel (incorrectly) plots ordered pairs at (-2,0) and (2,0).

Section 2-5 Exponential Functions

Example 1: Graphing Exponential Functions

We first enter *x*-values starting at -2 and increasing by 1.

Enter: The value **-2** in cell **A2**.
Enter: Formula **=A2+1** in cell **A3**.
Copy: Contents of cell **A3** to cells **A4:A6**.

Alternative: Enter values -2 and -1 in cells **A2** and **A3** respectively. Highlight both cells and hold down the left mouse button while over the lower right corner of cell **A3**, Then, drag the mouse down to cell **A6**.

Enter: Formula **=(1/2)*4^A2** in cell **B2**.
Copy: Contents of cell **B2** to cells **B3:B6**.

Good Habit: Hold down the left mouse button while over the lower right corner of cell **A2**, and drag the mouse down to cell **A6**. Do the same for column **B**.

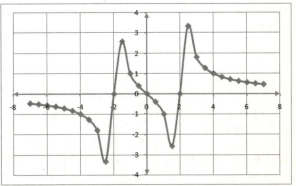

Select: Cells **A2:B6**

Select: On the **Insert** tab, select **Scatter** on the **Charts** menu. Then, select **Scatter with Smooth Lines and Markers**.

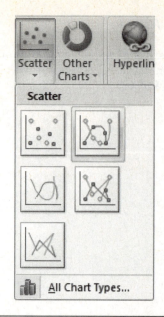

Select: **Design, Layout,** or **Format** tab under the **Chart Tools** menu. Use the submenus to format the output as desired.

For example,

Select: **Layout** tab and click on **Legend** on the **Labels** submenu. Then, select **None** to turn off the legend.

Select: **Layout** tab and click on **Data Labels** on the **Labels** submenu. Then, select **More Data Label Options** to add labels for the ordered pairs plotted.

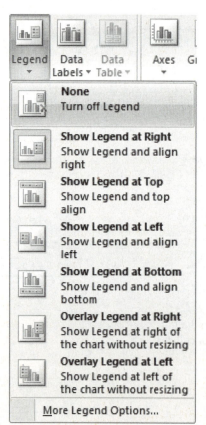

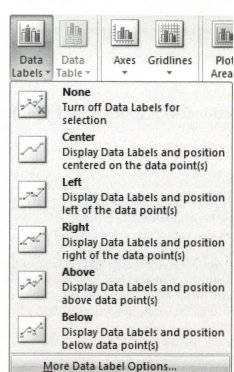

In the **Format Data Labels** window,
Select: **X Value**
Select: **Y Value**
Select: Label Position
Above
Click: **Close**

Using the **Chart Tools**, output can be formatted to look however you wish. One possibility is shown at the right.

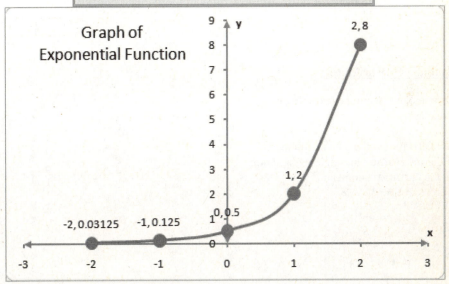

Example 2: Exponential Growth

Enter: The value **0.6** in cell **A2.**
Enter: Formula **=25*EXP(1.386*A2)** in cell **B2.**

	A	B	C
1	t	N(t)	
2		0.6	=25*EXP(1.386*A2)

Enter: The value **3.5** in cell **A2.**

	A	B
1	t	N(t)
2	3.5	3196.70485

Note: The exponential function is but one built-in Excel function. To access these, on the **Formulas** tab, select **Math & Trig**.

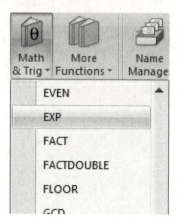

Example 3: Exponential Decay

Enter: The value **15000** in cell **A2.**
Enter: Formula **=500*EXP(-0.000124*A2)** in cell **B2.**

	A	B	C	D
1	t	N(t)		
2	15000	=500*EXP(-0.000124*A2)		

Enter: The value **45000** in cell **A2.**

	A	B
1	t	N(t)
2	45000	1.88628276

Example 4: Depreciation

Copy the data from the text into the worksheet.

Note: To format the data in column **B**, under the **Home** tab, select the **$** option on the **Number** submenu.

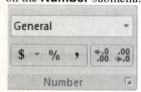

	A	B
1	x	Value ($)
2	1	$ 12,575
3	2	$ 9,455
4	3	$ 8,115
5	4	$ 6,845
6	5	$ 5,225
7	6	$ 4,485

Select: Cells **A2:B7**

Select: On the **Insert** tab, select **Scatter**
on the **Charts** menu. Then, select
Scatter with only Markers.

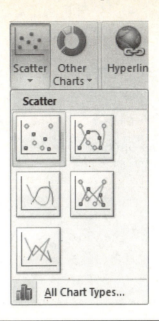

Select: **Design, Layout,** or **Format**
tab under the **Chart Tools** menu. Use
the submenus to format the output as
desired.
Using the **Chart Tools**, output can be
formatted to look however you wish.
One possibility is shown at the right.

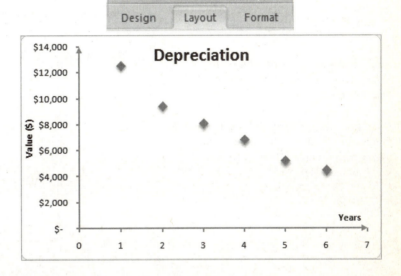

Select: **Layout** tab under the **Chart Tools** menu.

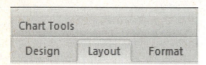

Select: **More Trendline Options** under the **Trendline** submenu.

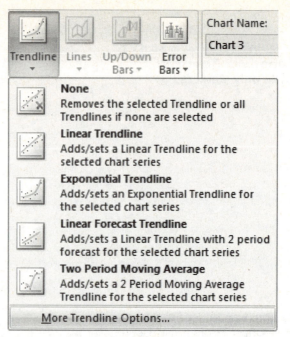

In the **Format Trendline** window,
Select: **Exponential** Trend/Regression Type
Enter: **Forecast: 1 periods** (optional)
Select: **Display Equation on chart**
Click: **Close**

Note: Forecasting 1 period forward graphically estimates the values of the data 1 period in the future (at time $x = 7$ years in this case).

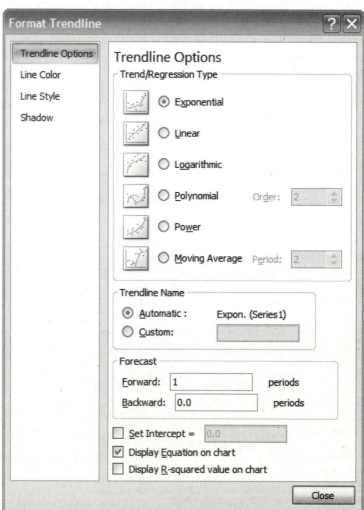

A regression line (an exponential model for the data) will appear on the chart.

Here, x represents the value of the independent variable and y represents the value of the dependent variable (Value in dollars).

It is important to note that Excel does not generate an exponential model of the form $y = ab^x$. Instead, Excel returns a function of the form $y = ae^{bx}$. Writing this as $y = a\left(e^b\right)^x$ gives the result in the previous format

$$y = 14910\left(e^{-0.203}\right)^x = 14910(0.816)^x.$$

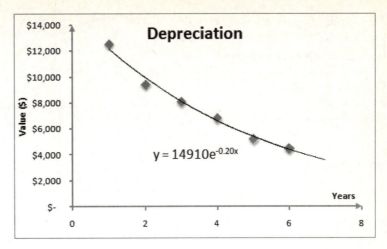

Example 5: Compound Growth

Enter: The value **1000** in cell **A2**.
Enter: The value **0.1** in cell **B2**.
Enter: The value **12** in cell **C2**.
Enter: The value **10** in cell **D2**.
Enter: Formula
=A2*(1+B2/C2)^(C2*D2) in cell **E2**.

	A	B	C	D	E
1	P	r	m	t	A
2	1000	0.1	12	10	=A2*(1+B2/

Section 2-6 Logarithmic Functions

Example 3: Solutions of the Equation $y = \log_b x$

Logarithm functions are built-in Excel functions. To access these, on the **Formulas** tab, select **Math & Trig**.

Page Layout	Formulas	Data

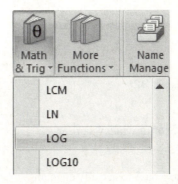

After highlighting the cell in which you would like to put $y = \log_4 16$, select **LOG** under the **Math & Trig** submenu.

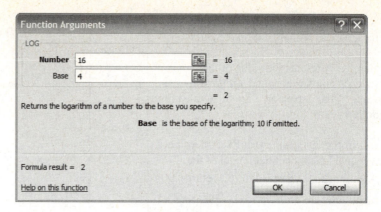

Enter: Number **16**
Enter: Base **4**
Click: **OK**

The resulting value, **2**, will be placed in the highlighted cell.

Remark: Alternatively, to compute $y = \log_4 16$,
Enter: Formula **=LOG(16,4)** in cell **A1**.

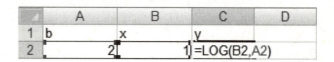

To solve $\log_2 x = -3$ for x,

Enter: The value **2** for the base b in cell **A2**.
Enter: Any positive value (such as **1**) in cell **B2**.
Enter: Formula **=LOG(B2,A2)** in cell **C2**.

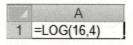

Remark: Alternatively, highlight cell **C2**, select **LOG** under the **Math & Trig** submenu.
Select: Cell **B2** for the Number
Select: Cell **A2** for the Base

Select: On the **Data** tab, select **Goal Seek** under the **What-If-Analysis** menu.

Enter: Set cell: **C2**
Enter: To value: **-3**
Enter: By changing cell: **B2**
Click: **OK**

When **Goal Seek Status** window appears,
Click: **OK**

Good Habit: Use the mouse to actually click on cells **C2** and **B2** for "Set cell:" and "By changing cell:" rather than typing the actual cell locations.

A "solution" will appear in cell **B2**. Note that this solution (0.12500117) is not exact.

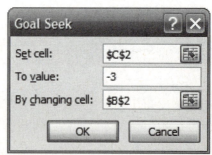

To solve $\log_b 100 = 2$ for b,

Enter: A guess for the value of b in cell **A2** (e.g. value $b = 3$).
Enter: The value **100** in cell **B2**.
Enter: Formula **=LOG(B2,A2)** in cell **C2**.

Remark: Alternatively, highlight cell **C2** and select **LOG** under the **Math & Trig** submenu.

Select: On the **Data** tab, select **Goal Seek** under the **What-If-Analysis** menu.

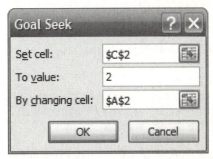

Enter: Set cell: **C2**
Enter: To value: **2**
Enter: By changing cell: **A2**
Click: **OK**

When **Goal Seek Status** window appears,
Click: **OK**

Good Habit: Use the mouse to actually click on cells **C2** and **A2** for "Set cell:" and "By changing cell:" rather than typing the actual cell locations.

A "solution" will appear in cell **A2**. Note that this solution (9.997167) is not exact.

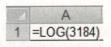

Example 7: Calculator Evaluation of Logarithms

To compute $\log 3{,}184$,

Enter: Formula **=LOG(3184)** in cell **A1**.

To access the natural logarithm built-in function of Excel, highlight cell **A1** and select **LN** under the **Math & Trig** submenu.

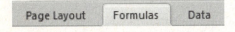

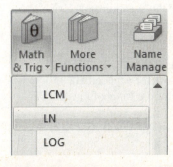

To compute $\ln 0.000349$,
Enter: Formula **=LN(0.000349)** in cell **A1**.

To compute log(−3.24),
Enter: Formula **=LOG(-3.24)** in cell **A1**.

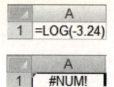

	A
1	=LOG(-3.24)

Since -3.24 is not in the domain of the log function, an error message occurs in cell **A1**.

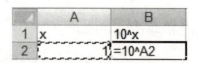

	A
1	#NUM!

Example 9: Solving Exponential Equations

To solve $10^x = 2$ for *x*,
Enter: A guess for the value of *x* in cell **A2** (e.g. value *x*=**1**).
Enter: Formula **=10^A2** in cell **B2**.
Select: On the **Data** tab, select **Goal Seek** under the **What-If-Analysis** menu.

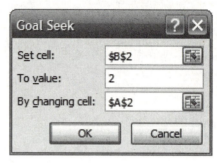

	A	B
1	x	10^x
2	1	=10^A2

Goal Seek

Set cell: $B$2

To value: 2

By changing cell: $A$2

Enter: Set cell: **B2**
Enter: To value: **2**
Enter: By changing cell: **A2**
Click: **OK**

When **Goal Seek Status** window appears,
Click: **OK**

Good Habit: Use the mouse to actually click on cells **B2** and **A2** for "Set cell:" and "By changing cell:" rather than typing the actual cell locations.

A "solution" will appear in cell **A2**. Note that this solution (0.30104498) is not exact (the exact solution has value log 2). Solving this type of problem algebraically (using logarithms) provides an exact solution and is advised.

	A	B
1	x	10^x
2	0.30104498	2.00006901

Example 10: Doubling Time for an Investment

To solve $2P = P\left(1+\dfrac{r}{m}\right)^{t \cdot m}$ for *t*, divide by *P* on both sides of the equation and solve

$2 = \left(1+\dfrac{r}{m}\right)^{t \cdot m}$ for *t*.

Enter: The value **0.2** for *r* in cell **A2**
Enter: The value **1** for *m* in cell **B2**
Enter: A guess for the value of *t* in cell **C2** (e.g. value *t* = **5**).
Enter: Formula **=(1+A2/B2)^(C2*B2)** in cell **D2**.

	A	B	C	D	E
1	r	m	t	A	
2	0.2	1	5	=(1+A2/B2)^(C2*B2)	

Select: On the **Data** tab, select **Goal Seek**
under the **What-If-Analysis** menu.

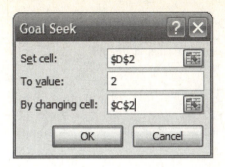

Enter: Set cell: **D2**
Enter: To value: **2**
Enter: By changing cell: **C2**
Click: **OK**

When **Goal Seek Status** window appears,
Click: **OK**

A "solution" will appear in cell **C2.** Note that
this solution (3.801806) is not exact.

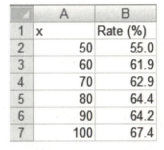

	A	B	C	D
1	r	m	t	A
2	0.2	1	3.801806	2.000008

Example 11: Home Ownership Rates

Copy the data from the text into the
worksheet.

	A	B
1	x	Rate (%)
2	50	55.0
3	60	61.9
4	70	62.9
5	80	64.4
6	90	64.2
7	100	67.4

Select: Cells **A2:B7**

Select: On the **Insert** tab, select
Scatter on the **Charts** menu. Then,
select **Scatter with only Markers**.

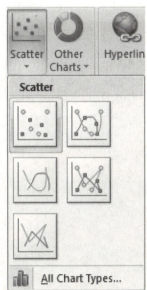

Select: **Design, Layout,** or **Format** tab under the **Chart Tools** menu. Use the submenus to format the output as desired.

Sample output is shown at the right.

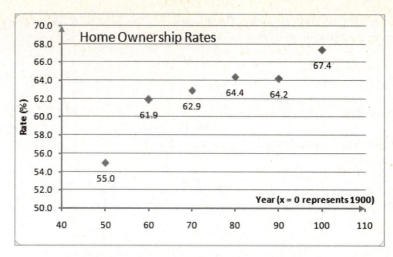

Select: **Layout** tab under the **Chart Tools** menu.

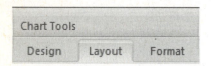

Select: **More Trendline Options** under the **Trendline** submenu.

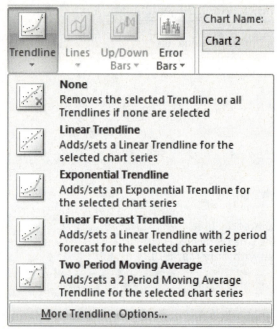

On the **Format Trendline** menu,
Select: **Logarithmic**
Trend/Regression Type
Select: **Display Equation on chart**
Click: **Close**

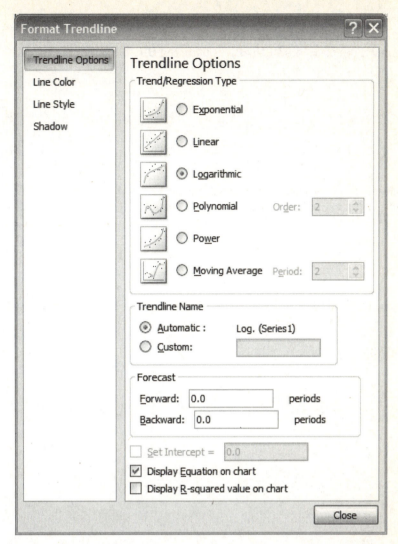

A regression line (a logarithmic model for the data) will appear on the chart.

Here, *x* represents the value of the independent variable (number of years after 1900) and y represents the value of the dependent variable (home ownership rate as a percentage).

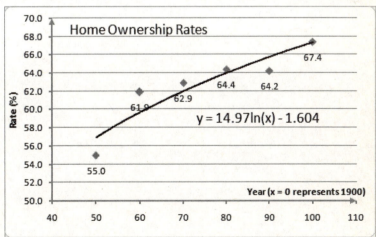

To predict the home ownership rate in 2015,
Enter: Value **115** in cell **A8**
Enter: Formula **=14.97*LN(A8)-1.604** in cell **B8**

⊿	A	B	C
1	x	Rate (%)	
2	50	55.0	
3	60	61.9	
4	70	62.9	
5	80	64.4	
6	90	64.2	
7	100	67.4	
8	115	=14.97*LN(A8)-1.604	

Conclude that this model estimates the home
ownership rate to be 69.4% in 2015.

⊿	A	B
1	x	Rate (%)
2	50	55.0
3	60	61.9
4	70	62.9
5	80	64.4
6	90	64.2
7	100	67.4
8	115	69.4

Chapter 3

Section 3-1 Simple Interest

Example 1: Total Amount Due on a Loan

Enter: **800** in cell **A2**
Enter: **0.09** in cell **B2**
Enter: Formula **=4/12** in cell **C2**
Enter: Formula **=A2*(1+B2*C2)** in cell **D2**

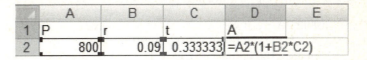

	A	B	C	D	E
1	P	r	t	A	
2	800	0.09	0.333333	=A2*(1+B2*C2)	

Optional:
Select: Cell **D2**
On the **Home** tab, select **$** (Accounting) on the **Number** menu to format the contents of the cell.

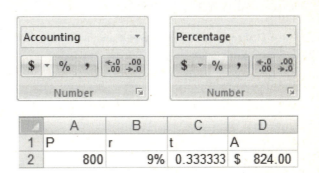

Optional:
Select: Cell **B2**
On the **Home** tab, select **%** (Percentage) on the **Number** menu to format the contents of the cell.

	A	B	C	D
1	P	r	t	A
2	800	9%	0.333333	$ 824.00

Example 2: Present Value of an Investment

Enter: Any value (e.g. **$4000**) in cell **A2**
Enter: **0.1** (10%) in cell **B2**
Enter: Formula **=9/12** in cell **C2**
Enter: Formula **=A2*(1+B2*C2)** in cell **D2**

Select: On the **Data** tab, select **Goal Seek** under the **What-If-Analysis** menu.

	A	B	C	D	E
1	P	r	t	A	
2	4000	0.1	0.75	=A2*(1+B2*C2)	

Enter: Set cell: **D2**
Enter: To value: **5000**
Enter: By changing cell: **A2**
Click: **OK**

When **Goal Seek Status** window appears,
Click: **OK**

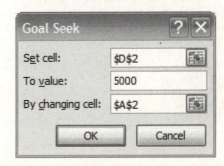

Good Habit: Use the mouse to actually click on cells **D2** and **A2** for "Set cell:" and "By changing cell:" rather than typing the actual cell locations.

53

A "solution" ($4651.16) will appear in cell **A2.**

	A	B	C	D
1	P	r	t	A
2	$4,651.16	10%	0.75	$5,000.00

Here, the data has been formatted using the **Number** menu on the **Home** tab.

Example 3: Interest Rate Earned on a Note

Enter: **$9893.78** in cell **A2**
Enter: Any value (e.g. **0.05** or 5%) in cell **B2**
Enter: Formula **=180/360** in cell **C2**
Enter: Formula **=A2*(1+B2*C2)** in cell **D2**

	A	B	C	D
1	P	r	t	A
2	9893.78	0.05	=180/360	

Select: On the **Data** tab, select **Goal Seek** under the **What-If-Analysis** menu.

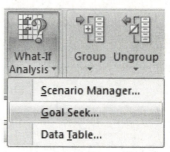

Enter: Set cell: **D2**
Enter: To value: **10000**
Enter: By changing cell: **B2**
Click: **OK**

When **Goal Seek Status** window appears, Click: **OK**

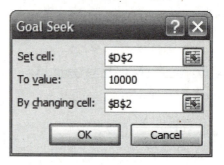

A "solution" (0.021472) will appear in cell **B2.**

	A	B	C	D
1	P	r	t	A
2	9893.78	0.021472	0.5	10000

Optional:
Select: **Cell B2**
On the **Home** tab, select **%** (Percentage) on the **Number** menu to format the contents of the cell. Use the **Increase Decimal** button (highlighted below) to increase the number of decimal places.

	A	B	C	D
1	P	r	t	A
2	$9,893.78	2.147%	0.5	$10,000.00

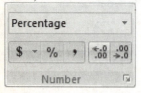

Example 5: Interest on an Investment

Copy the data from the text into the worksheet.
Note: On the **Home** tab, select the **Number** menu to format the contents (e.g., format dollars as **Currency**).

	A	B	C
1	Transaction Minimum	Fixed Commission	Variable Commission
2	$0.00	$29.00	1.6%
3	$2,500.00	$49.00	0.8%
4	$10,000.00	$99.00	0.3%

Enter: Value **50** in cell **A7**
Enter: Value **$47.52** in cell **B7**
Enter: Formula **=A7*B7** in cell **C7**
Enter: Formula **=IF(C7<A3,B2+C2*C7,IF(C7<A4,B3+C3*C7,B4+C4*C7))** in cell **D7**
Enter: Formula **=C7+D7** in cell **E7**
Note: On the **Home** tab, select the **Number** menu to format the contents.

	A	B	C	D	E	F	G	H
1	Transaction Minimum	Fixed Commission	Variable Commission					
2	$0.00	$29.00	1.6%					
3	$2,500.00	$49.00	0.8%					
4	$10,000.00	$99.00	0.3%					
5								
6	Shares bought	Cost/Share	Principal	Commission	Total Investment			
7	50	$47.52	$2,376.00	=IF(C7<$A$3,B2+C2*C7,IF(C7<$A$4,B3+C3*C7,B4+C4*C7))				

	A	B	C	D	E
1	Transaction Minimum	Fixed Commission	Variable Commission		
2	$0.00	$29.00	1.6%		
3	$2,500.00	$49.00	0.8%		
4	$10,000.00	$99.00	0.3%		
5					
6	Shares bought	Cost/Share	Principal	Commission	Total Investment
7	50	$47.52	$2,376.00	$67.02	=C7+D7

Enter: Value **50** in cell **A10**
Enter: Value **$52.19** in cell **B10**
Enter: Formula **=A10*B10** in cell **C10**
Enter: Formula **=IF(C10<A3,B2+C2*C10,IF(C10<A4,B3+C3*C10,B4+C4*C10))** in cell **D10**
Enter: Formula **=C10-D10** in cell **E10**

	A	B	C	D	E
1	Transaction Minimum	Fixed Commission	Variable Commission		
2	$0.00	$29.00	1.6%		
3	$2,500.00	$49.00	0.8%		
4	$10,000.00	$99.00	0.3%		
5					
6	Shares bought	Cost/Share	Principal	Commission	Total Investment
7	50	$47.52	$2,376.00	$67.02	$2,443.02
8					
9	Shares sold	Cost/Share	Principal	Commission	Total Return
10	50	$52.19	$2,609.50	$69.88	=C10-D10

Enter: Formula **=E7** in cell **A13** (Do you see why?)
Enter: Any value (e.g. **5%**) in cell **B13**
Enter: Formula **=200/360** in cell **C13**
Enter: Formula **=A13*(1+B13*C13)** in cell **D13**
Enter: Formula **D13-E10** in cell **E13**

	A	B	C	D	E
1	Transaction Minimum	Fixed Commission	Variable Commission		
2	$0.00	$29.00	1.6%		
3	$2,500.00	$49.00	0.8%		
4	$10,000.00	$99.00	0.3%		
5					
6	Shares bought	Cost/Share	Principal	Commission	Total Investment
7	50	$47.52	$2,376.00	$67.02	$2,443.02
8					
9	Shares sold	Cost/Share	Principal	Commission	Total Return
10	50	$52.19	$2,609.50	$69.88	$2,539.62
11					
12	P	r	t	A	A - Total Return
13	$2,443.02	5.000%	0.555555556	=A13*(1+B13*C13)	

Select: On the **Data** tab, select **Goal Seek** under the **What-If-Analysis** menu.

Enter: Set cell: **E13**
Enter: To value: **0**
Enter: By changing cell: **B13**
Click: **OK**

When **Goal Seek Status** window appears, Click: **OK**

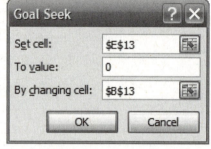

A "solution" (7.118%) will appear in cell **B13**.

Note the beauty of this "dynamic" construction is that it allows several different parameters (such as the fixed commission or the cost/share amount) to be changed and automatically updates the result.

	A	B	C	D	E
1	Transaction Minimum	Fixed Commission	Variable Commission		
2	$0.00	$29.00	1.6%		
3	$2,500.00	$49.00	0.8%		
4	$10,000.00	$99.00	0.3%		
5					
6	Shares bought	Cost/Share	Principal	Commission	Total Investment
7	50	$47.52	$2,376.00	$67.02	$2,443.02
8					
9	Shares sold	Cost/Share	Principal	Commission	Total Return
10	50	$52.19	$2,609.50	$69.88	$2,539.62
11					
12	P	r	t	A	A - Total Return
13	$2,443.02	7.118%	0.555555556	$2,539.62	$0.00

Section 3-2 Compound and Continuous Compound Interest

Example 1: Comparing Interest for Various Compounding Periods

Compounding annually:
Enter: Value $**1000** in cell **A2**
Enter: Value **0.08** in cell **B2**
Enter: Value **5** in cell **C2**
Enter: Formula **=A2*(1+B2)^C2** in cell **D2**

	A	B	C	D	E
1	P	i	n	A	
2	$1,000.00	0.08	5	=A2*(1+B2)^C2	

Compounding semiannually:
Enter: Formula **=0.08/2** in cell **B2**
Enter: Formula **=2*5** in cell **C2**

	A	B	C	D
1	P	i	n	A
2	$1,000.00	0.04	=2*5	$1,480.24

Compounding quarterly:
Enter: Formula **=0.08/4** in cell **B2**
Enter: Formula **=4*5** in cell **C2**

	A	B	C	D
1	P	i	n	A
2	$1,000.00	=0.08/4	20	$1,485.95

Compounding monthly:
Enter: Formula =**0.08/12** in cell **B2**
Enter: Formula =**12*5** in cell **C2**

	A	B	C	D
1	P	i	n	A
2	$ 1,000.00	0.006667	60	$ 1,489.85

Example 2: Compounding Daily and Continuously

Enter: Value **$5000** in cell **A2**.
Enter: Value **0.08** in cell **B2**.
Enter: Value **365** in cell **C2**.
Enter: Value **2** in cell **D2**.

Enter: Formula
=**A2*(1+B2/C2)^(C2*D2)** in cell **E2**.

	A	B	C	D	E	F
1	P	r	m	t	A	
2	$ 5,000.00	0.08	365	2	=A2*(1+B2/C2)^(C2*D2)	

	A	B	C	D	E
1	P	r	m	t	A
2	$ 5,000.00	0.08	365	2	$ 5,867.45

Enter: Value **$5000** in cell **A2**.
Enter: Value **0.08** in cell **B2**.
Enter: Value **2** in cell **C2**.

Enter: Formula =**A2*EXP(B2*C2)** in
cell **D2**.

	A	B	C	D	E
1	P	r	t	A	
2	$5,000.00	0.08	2	=A2*EXP(B2*C2)	

	A	B	C	D
1	P	r	t	A
2	$5,000.00	0.08	2	$5,867.55

Example 3: Finding Present Value

Interest compounded quarterly:
Enter: Any value (e.g. **$6000**) in cell
A2.
Enter: Formula =**0.10/4 (2.5%)** in
cell **B2**.
Enter: Formula =**4*5** in cell **C2**.
Enter: Formula =**A2*(1+B2)^C2** in
cell **D2**.
Select: On the **Data** tab, select **Goal
Seek** under the **What-If-Analysis**
menu.

	A	B	C	D	E
1	P	i	n	A	
2	$6,000.00	0.025	20	=A2*(1+B2)^C2	

Enter: Set cell: **D2**
Enter: To value: **8000**
Enter: By changing cell: **A2**
Click: **OK**

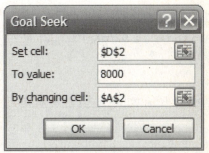

Goal Seek Status window will
appear.
Click: **OK**
The required present value *P* will
appear in cell **A2**.

	A	B	C	D
1	P	i	n	A
2	$4,882.17	0.025	20	$8,000.00

A second approach is one in which simple algebra first solves for the present value in terms of *A*:

$$P = \frac{A}{(1+i)^n}$$

Enter: Value **$8000** in cell **A2**

Enter: Formula = **A2/(1+B2)^C2** in cell **D2**.

	A	B	C	D	E
1	A	i		n	P
2	$8,000.00	0.025		20	=A2/(1+B2)^C2

Interest compounded continuously:
Enter: Any value (e.g. **$6000**) in cell **A2**.
Enter: Value **0.10 (10%)** in cell **B2**.
Enter: Value **5** in cell **C2**.
Enter: Formula **=A2*EXP(B2*C2)** in cell **D2**.
Select: On the **Data** tab, select **Goal Seek** under the **What-If-Analysis** menu.

	A	B	C	D	E
1	P	r	t	A	
2	$6,000.00	0.1	5	=A2*EXP(B2*C2)	

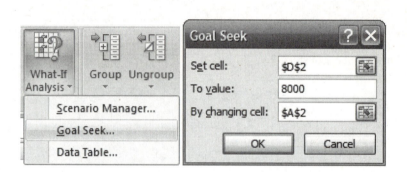

Enter: Set cell: **D2**
Enter: To value: **8000**
Enter: By changing cell: **A2**
Click: **OK**

Goal Seek Status window will appear.
Click: **OK**
The required present value will appear in cell **A2**.

	A	B	C	D
1	P	r	t	A
2	$4,852.25	0.1	5	$8,000.00

A second approach is one in which simple algebra first solves for the present value in terms of *A*: $P = \frac{A}{e^{rt}}$

Enter: Value **$8000** in cell **A2**

Enter: Formula = **A2/EXP(B2*C2)** in cell **D2**.

	A	B	C	D	E
1	A	r	t	P	
2	$8,000.00	0.1	5	=A2/EXP(B2*C2)	

Example 4: Computing Growth Rate

Interest compounded annually:
Enter: Present value **$10000** in cell **A2**.
Enter: Any value (e.g. **0.2** or **20%**) in cell **B2**.
Enter: Value **10** in cell **C2**.
Enter: Formula **=A2*(1+B2)^C2** in cell **D2**.

	A	B	C	D	E
1	P	r	t	A	
2	$10,000.00	0.2	10	=A2*(1+B2)^C2	

Select: On the **Data** tab, select **Goal Seek** under the **What-If-Analysis** menu.

Enter: Set cell: **D2**
Enter: To value: **126000**
Enter: By changing cell: **B2**
Click: **OK**

Goal Seek Status window will appear.
Click: **OK**
The growth rate (28.8359%) will appear in cell **B2**.

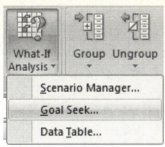
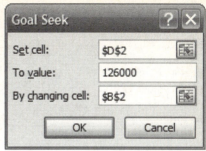

	A	B	C	D
1	P	r	t	A
2	$10,000.00	0.288359	10	$126,000.00

Interest compounded continuously:
Enter: Present value **$10000** in cell **A2**.
Enter: Any value (e.g. **0.2**) in cell **B2**.
Enter: Value **10** in cell **C2**.
Enter: Formula **=A2*EXP(B2*C2)** in cell **D2**.
Select: On the **Data** tab, select **Goal Seek** under the **What-If-Analysis** menu.

	A	B	C	D	E
1	P	r	t	A	
2	$10,000.00	0.2	10	=A2*EXP(B2*C2)	

Enter: Set cell: **D2**
Enter: To value: **126000**
Enter: By changing cell: **B2**
Click: **OK**

Goal Seek Status window will appear.
Click: **OK**
The growth rate (25.337%) will appear in cell **B2**.

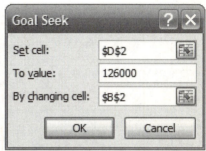

	A	B	C	D
1	P	r	t	A
2	$10,000.00	0.25337	10	$126,000.00

Example 5: Computing Growth Time

Enter: Value **$10000** in cell **A2**.
Enter: Formula **=0.09/12** in cell **B2**.
Enter: Any value (e.g. **12** months) in cell **C2**.
Enter: Formula **=A2*(1+B2)^C2** in cell **D2**.

	A	B	C	D
1	P	i	n	A
2	$10,000.00	=0.09/12	12	$10,938.07

Select: On the **Data** tab, select **Goal Seek** under the **What-If-Analysis** menu.

Enter: Set cell: **D2**
Enter: To value: **12000**
Enter: By changing cell: **C2**
Click: **OK**

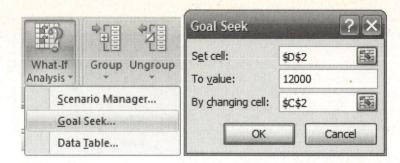

Goal Seek Status window will appear.
Click: **OK**
The growth time will appear in cell **C2.** Since the growth time must be a positive integer in this example, 24.4 is rounded up to 25 months.

	A	B	C	D
1	P	i	n	A
2	$10,000.00	0.0075	24.40059	$12,000.00

Example 6: Using APY to Compare Investments

Copy the data from the text into the worksheet.

	A	B	C
1	Bank	Rate	m
2	Advanta	4.93%	12
3	DeepGreen	4.95%	365
4	Charter One	4.97%	4
5	Liberty	4.94%	

Enter: Formula **=(1+B2/C2)^C2-1** in cell **D2**

	A	B	C	D	E
1	Bank	Rate	m	APY	
2	Advanta	4.93%	12	=(1+B2/C2)^C2-1	

Copy: Contents of cell **D2** to cells **D3:D4**

	A	B	C	D
1	Bank	Rate	m	APY
2	Advanta	4.93%	12	5.043%
3	DeepGreen	4.95%	365	5.074%
4	Charter One	4.97%	4	5.063%
5	Liberty	4.94%		

Good habit: Select cell **D2.** Hold mouse over right lower corner, click and hold the left button dragging down to cell **D4**

Enter: Formula **=EXP(B2)-1** in cell **D5**

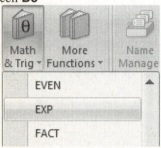

Note: The exponential function is but one built-in Excel function. To access these, on the **Formulas** tab, select **Math & Trig**.

	A	B	C	D
1	Bank	Rate	m	APY
2	Advanta	4.93%	12	5.043%
3	DeepGreen	4.95%	365	5.074%
4	Charter One	4.97%	4	5.063%
5	Liberty	4.94%		=EXP(B5)-1

Example 7: Computing the Annual Nominal Rate Given the APY

Enter: Any value (e.g. **0.05** or **5%**) in cell **A2**
Enter: Value **12** in cell **B2**
Enter: Formula **=(1+A2/B2)^B2-1** in cell **C2**

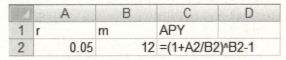

	A	B	C	D
1	r	m	APY	
2	0.05	12	=(1+A2/B2)^B2-1	

Select: On the **Data** tab, select **Goal Seek** under the **What-If-Analysis** menu.

Enter: Set cell: **C2**
Enter: To value: **0.075**
Enter: By changing cell: **A2**
Click: **OK**

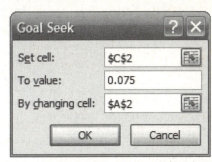

Goal Seek Status window will appear.
Click: **OK**

Remark: As always, remember that Excel can only produce approximate solutions (when it can find solutions at all).

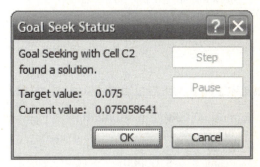

The (approximate) annual nominal rate *r* will appear in cell **A2**. Note that this rate actually produces an APY of about 7.5059%.

	A	B	C
1	r	m	APY
2	0.072594	12	0.075059

Section 3-3 Future Value of an Annuity; Sinking Funds

Example 1: Future Value of an Ordinary Annuity

Enter: Value **0.085** (or 8.5%) in cell **A2**.
Enter: Value **20** in cell **B2**.
Enter: Payment Value **$2000** in cell **C2**.

	A	B	C
1	i	n	PMT
2	0.085	20	2000

The future value function is a built-in Excel function. To access it, on the **Formulas** tab, select **Financial**.

Select: Cell **D2**
Select: **FV** from the **Financial Formulas**.

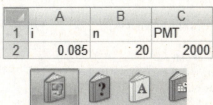

In **Function Arguments** window,

Enter: **Rate** (Select cell **A2**)
Enter: **Nper** (Select cell **B2**) [**Nper** is
short for number of periods.]
Enter: **Pmt** (Select cell **C2**)

Click: **OK**

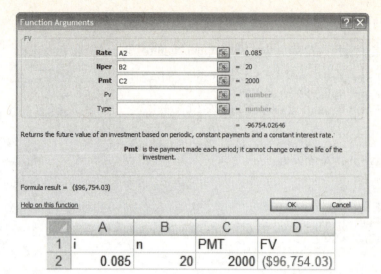

Remark: Output is negative (it is in red and
in parentheses). Making the PMT negative
will make FV positive. It's all a matter of
perspective.
Remark: One alternative path to arrive at
this point in computing is to
Enter: Formula **=FV(A2,B2,C2)** in cell
D2.

	A	B	C	D
1	i	n	PMT	FV
2	0.085	20	2000	($96,754.03)

Example 2: Computing the Payment for a Sinking Fund

Enter: Formula **=0.066/12** in cell **A2** (this
represents the monthly rate)
Enter: Value **=5*12** in cell **B2**
Enter: Any value (e.g. **$100000**) in cell **C2**

	A	B	C
1	i	n	PMT
2	0.5500%	60	$ 100,000.00

The future value function is a built-in Excel
function. To access the built-in Excel
function for future value, on the **Formulas**
tab, select **Financial**.

Select: Cell **D2**
Select: **FV** from the **Financial Formulas**.

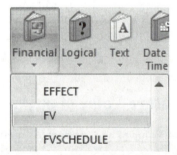

In **Function Arguments** window,

Enter: **Rate** (Select cell **A2**)
Enter: **Nper** (Select cell **B2**) [**Nper** is short
for number of periods.]
Enter: **Pmt** (Select cell **C2**)

Click: **OK**

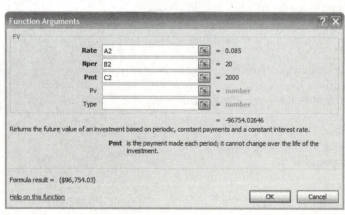

Select: On the **Data** tab, select **Goal Seek** under the **What-If-Analysis** menu.

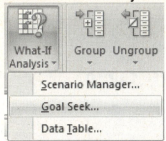

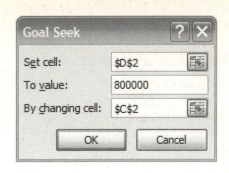

Enter: Set cell: **D2**
Enter: To value: **800000**
Enter: By changing cell: **C2**
Click: **OK**

Goal Seek Status window will appear.
Click: **OK**
The monthly payment ($11,290.42) will appear in cell **C2**. Note that this number is negative (this amount is paid into an account each month).

	A	B	C	D
1	i	n	PMT	FV
2	0.5500%	60	$ (11,290.42)	$800,000.00

Example 4: Approximating an Interest Rate

Enter: Any value (e.g. **0.005** or **0.5%**) in cell **A2**. *Note*: This value should be reasonably close to the actual monthly interest rate.
Enter: Value **360** in cell **B2**.
Enter: Value **-100** in cell **C2**.
Enter: Formula **=FV(A2,B2,C2)** in cell **D2**. (Or, on the **Formulas** tab, select **Financial** and the built-in **FV** function)

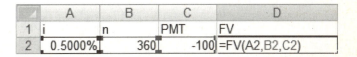

	A	B	C	D
1	i	n	PMT	FV
2	0.5000%	360	-100	=FV(A2,B2,C2)

Select: On the **Data** tab, select **Goal Seek** under the **What-If-Analysis** menu.

Enter: Set cell: **D2**
Enter: To value: **160000**
Enter: By changing cell: **A2**
Click: **OK**

Goal Seek Status window will appear.
Click: **OK**

The (approximate) monthly interest rate (0.6957%) will appear in cell **A2**. To convert this to an annual rate, multiply this number by 12 (to get about 8.35%).

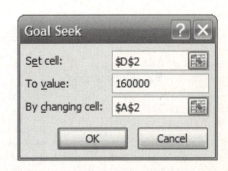

	A	B	C	D
1	i	n	PMT	FV
2	0.6957%	360	-100	$160,000.00

Section 3-4 Present Value of an Annuity; Amortization

<u>**Example 1: Present Value of an Annuity**</u>

Enter: Formula **=0.06/12** in cell **A2.**
Enter: Formula **=5*12** in cell **B2.**
Enter: Value **$200** in cell **C2.**

	A	B	C
1	i	n	PMT
2	0.5000%	60	$ 200.00

Select: Cell **D2**

The present value function is a built-in Excel function. To access it, on the **Formulas** tab, select **Financial**.

Select: **FV** from the **Financial Formulas**.

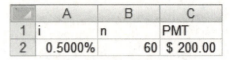

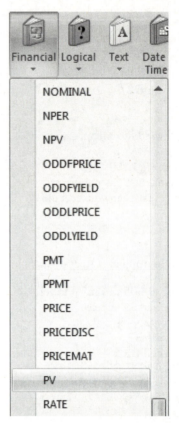

In **Function Arguments** window,

Enter: **Rate** (Select cell **A2**)
Enter: **Nper** (Select cell **B2**) [**Nper** is short for number of periods.]
Enter: **Pmt** (Select cell **C2**)

Click: **OK**

Alternatively, one can type directly in the cell:
Enter: Formula **=PV(A2,B2,C2)** in cell **D2.**

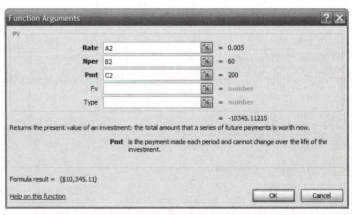

Note: Output is negative. Making the PMT negative will make PV positive. It's all a matter of perspective.

	A	B	C	D
1	i	n	PMT	PV
2	0.5000%	60	$ 200.00	($10,345.11)

Example 2: Retirement Planning

Enter: Value **0.065** or **6.5%** in cell **A2**.
Enter: Value **20** in cell **B2**.
Enter: Value **-25000** in cell **C2**.
Enter: Formula **=PV(A2,B2,C2)** in cell **D2**.

Enter: Formula **=A2** in cell **A5**.
Enter: Value **25** in cell **B5**.
Enter: Formula **=D2** in cell **C5**.

	A	B	C	D	E
1	i	n (withdrawals)	PMT (out)	PV	
2	6.5000%	20	$ (25,000.00)	=PV(A2,B2,C2)	

	A	B	C	D
1	i	n (withdrawals)	PMT (out)	PV
2	6.5000%	20	$ (25,000.00)	$275,462.68
3				
4	i	n (deposits)	FV	PMT (in)
5	6.5000%	25	$275,462.68	

Select: Cell **D5**

The payment function is a built-in Excel function. To access it, on the **Formulas** tab, select **Financial**.

Select: **PMT** from the **Financial Formulas**.

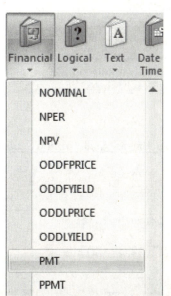

In **Function Arguments** window,

Enter: **Rate** (Select cell **A5**)
Enter: **Nper** (Select cell **B5**) [**Nper** is short for number of periods.]
Enter: **Pv** of **0**
Enter: **Fv** (Select cell **C5**)

Click: **OK**

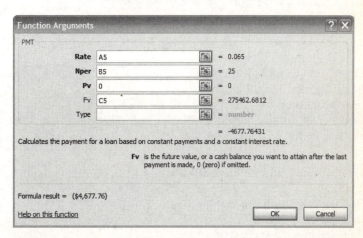

Alternatively, one can type directly in the cell:
Enter: Formula **=PMT(A5,B5,0,C5)** in cell **D5**.

	A	B	C	D
1	i	n (withdrawals)	PMT (out)	PV
2	6.5000%	20	$ (25,000.00)	$275,462.68
3				
4	i	n (deposits)	FV	PMT (in)
5	6.5000%	25	$275,462.68	($4,677.76)

Example 3: Monthly Payment and Total Interest on an Amortized Debt

Enter: Value **0.015** (1.5%) in cell **A2**.
Enter: Value **18** in cell **B2**.
Enter: Value **800** in cell **C2**.
Enter: Formula **=PMT(A2,B2,C2)** in cell **D2**.
Note: The value of PMT is negative.

	A	B	C	D	E
1.	i	n	PV	PMT	Total Interest
2	1.5000%	18	$ 800.00	=PMT(A2,B2,C2)	

Enter: Formula **=B2*(-D2)-C2** in cell **E5**.

	A	B	C	D	E
1	i	n	PV	PMT	Total Interest
2	1.5000%	18	$ 800.00	($51.04)	=B2*(-D2)-C2

Note: Because $51.04 has been rounded (down) to the nearest penny, the total interest paid is actually less than what it is in theory. In theory, the total interest (as shown in Excel) is roughly $118.80. However, since each payment was rounded down to the nearest penny, the amount of interest paid via payments of size $51.04 is actually less.

Example 4: Constructing an Amortization Schedule

Enter: Value **0.01** (1%) in cell **A2**.
Enter: Value **6** in cell **B2**.

	A	B	C	D
1	i	n	PV	PMT
2	1.0000%	6	$ (500.00)	=PMT(A2,B2,C2)

Enter: Value **-500** (amount borrowed) in cell **C2**.
Enter: Formula **=PMT(A2,B2,C2)** in cell **D2**.
Enter: Value **0** in cell **A5**
Enter: Formula **=-C2** in cell **E5**

	A	B	C	D	E
1	i	n	PV	PMT	
2	1.0000%	6	$ (500.00)	$86.27	
3					
4	Payment Number	Payment	Interest	Unpaid Balance Reduction	Unpaid Balance
5	0				=-C2

Enter: Formula **=A5+1** in cell **A6**
Enter: Formula **=$D$2** (an absolute reference) in cell **B6**
Remark: An absolute reference will refer to the same cell even upon copying.

	A	B	C	D	E
1	i	n	PV	PMT	
2	1.0000%	6	$ (500.00)	$86.27	
3					
4	Payment Number	Payment	Interest	Unpaid Balance Reduction	Unpaid Balance
5	0				$500.00
6	1	=$D$2			

Enter: Formula **=$A$2*E5** in cell **C6**

Remark: Once again, an absolute reference (like **$A$2**) will refer to the same cell even upon copying.

	A	B	C	D	E
1	i	n	PV	PMT	
2	1.0000%	6	$ (500.00)	$86.27	
3					
4	Payment Number	Payment	Interest	Unpaid Balance Reduction	Unpaid Balance
5	0				$500.00
6	1	$86.27	=$A$2*E5		

Enter: Formula **=B6-C6** in cell **D6**
Enter: Formula **=E5-D6** in cell **E6**

	A	B	C	D	E
1	i	n	PV	PMT	
2	1.0000%	6	$ (500.00)	$86.27	
3					
4	Payment Number	Payment	Interest	Unpaid Balance Reduction	Unpaid Balance
5	0				$500.00
6	1	$86.27	$5.00	$81.27	=E5-D6

Select: Cells **A6:E6**
Copy: Values of cells **A6:E6** to cells
A7:E11

To accomplish this, highlight cells
A6:E6. Then, drag lower right hand
corner of highlighted area down using
mouse.

Note: The amounts given in columns D
and E are slightly different than those
given in the text due to rounding. When
Excel does calculations, even though
only two digits after the decimal place
are shown (in this table), more digits are
being used.

	A	B	C	D	E
1	i	n	PV	PMT	
2	1.0000%	6	$ (500.00)	$86.27	
3					
4	Payment Number	Payment	Interest	Unpaid Balance Reduction	Unpaid Balance
5	0				$500.00
6	1	$86.27	$5.00	$81.27	$418.73
7	2	$86.27	$4.19	$82.09	$336.64
8	3	$86.27	$3.37	$82.91	$253.73
9	4	$86.27	$2.54	$83.74	$169.99
10	5	$86.27	$1.70	$84.57	$85.42
11	6	$86.27	$0.85	$85.42	$0.00

Chapter 4

Section 4-1 Review: Systems of Linear Equations in Two Variables

<u>**Example 7: Supply and Demand**</u>

The **Solver** Add-in for Excel can help analyze systems of equations. *If this Add-in has not been installed before*, you will need to install it.

Select: Large, round **Office Button**
Select: **Excel Options** button at bottom

In the **Excel Options** window,

Select: **Add-Ins** on the left menu
Select: **Solver Add-in**
Click: **Go…**

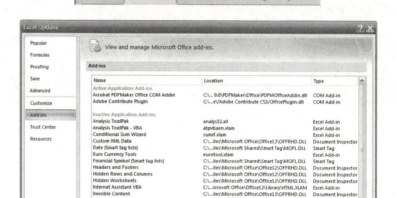

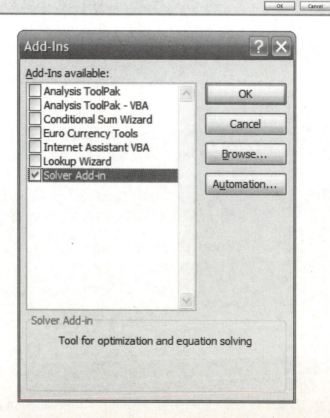

In the **Add-Ins** window,

Select: **Solver Add-in**
Click: **OK**

In the **Microsoft Office Excel** window,

Select: **Yes**

Enter: Any value (e.g. **0**) in cell **A2**
Enter: Any value (e.g. **0**) in cell **B2**
Enter: Formula **=A2** in cell **A5** (representing the left hand side of price-demand equation)
Enter: Formula **=-0.2*B2+4** in cell **B5** (representing the right hand side of price-demand equation)
Enter: Formula **=A2** in cell **A6** (representing the left hand side of price-supply equation)
Enter: Formula **=0.07*B2+0.76** in cell **B6** (representing the right hand side of price-supply equation)
Select: **Data** tab

	A	B	C
1	p	q	
2		0	0
3			
4	LHS	RHS	
5		0	4
6		0	=0.07*B2+0.76

| Formulas | Data | Review |

Select: **Solver**

Enter: **Select Target Cell: A2**
Select: **Equal To: Max (or Min!)**
Enter: **By Changing Cells: A2, B2**
Click: Subject to the Constraints: **Add**

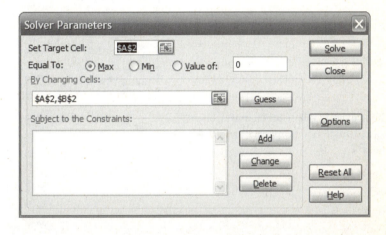

Enter: Cell Reference: **A5**
Select: **=**
Enter: Constraint: **B5**
Click: **Add**

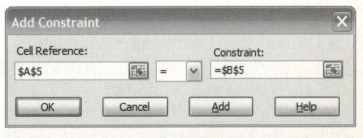

Enter: Cell Reference: **A6**
Select: **=**
Enter: Constraint: **B6**
Click: **OK**

Click: **Solve**

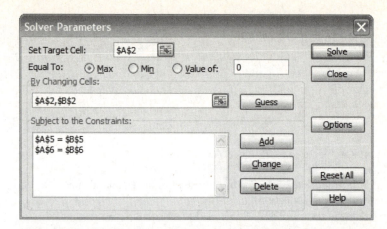

Select: **Keep Solver Solution**
Click: **OK**

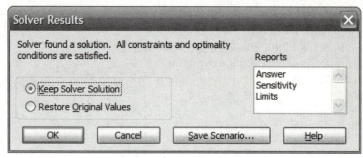

Since exactly one solution exists, an approximation to it appears in cells **A2** and **B2**.

	A	B
1	p	q
2	1.6	12
3		
4	LHS	RHS
5	1.6	1.6
6	1.6	1.6

Section 4-2 Systems of Linear Equations and Augmented Matrices

Example 1: Solving a System Using Augmented Matrix Methods

[Perform row operation $R_1 \leftrightarrow R_2$]
Enter: Contents of augmented matrix in cells **A1:C2**
Enter: Formula **=A2** in cell **A4** (this is the first step in switching the two rows)
Copy: Contents of cell **A4** to cells **B4:C4**
Good Habit: Drag bottom right corner of cell **A4** to cell **C4** using mouse.
Enter: Formula **=A1** in cell **A5**
Copy: Contents of cell **A5** to cells **B5:C5** (this completes the switching of the two rows)
Remark: Remember, copying the contents can be done by holding the left mouse button down while dragging the right lower corner of cell **A5** to cell **C5**.

	A	B	C
1	3	4	1
2	1	-2	7
3			
4	=A2		

	A	B	C
1	3	4	1
2	1	-2	7
3			
4	1	-2	7
5	3	4	1

[Perform row operation $(-3)R_1 + R_2 \to R_2$]

First, recopy row 1.

Enter: Formula **=A4** in cell **A7**

Copy: Contents of cell **A7** to cells **B7:C7**

Then, modify row 2.

Enter: Formula **=(-3)*A4+A5** in cell **A8**

Copy: Contents of cell **A8** to cells **B8:C8**

	A	B	C
1	3	4	1
2	1	-2	7
3			
4	1	-2	7
5	3	4	1
6			
7	1	-2	7
8	=(-3)*A4+A5		

[Perform row operation $\frac{1}{10}R_2 \to R_2$]

First, recopy row 1.

Enter: Formula **=A7** in cell **A10**

Copy: Contents of cell **A10** to cells **B10:C10**

Then, modify row 2.

Enter: Formula **=1/10*A8** in cell **A11**

Copy: Contents of cell **A11** to cells **B11:C11**

	A	B	C
7	1	-2	7
8	0	10	-20
9			
10	1	-2	7
11	0	=1/10*B8	

[Perform row operation $2R_2 + R_1 \to R_1$]

First, modify row 1.

Enter: Formula **=2*A11+A10** in cell **A13**

Copy: Contents of cell **A13** to cells **B13:C13**

Then, recopy row 2.

Enter: Formula **=A11** in cell **A14**

Copy: Contents of cell **A14** to cells **B14:C14**

The reduced augmented matrix can be found in cells **A13:C14**.

	A	B	C
10	1	-2	7
11	0	1	-2
12			
13	=2*A11+A10		

	A	B	C
13	1	0	3
14	0	1	-2

Example 3: Solving a System Using Augmented Matrix Methods

[Perform row operations $\frac{1}{2}R_1 \to R_1$ and $\frac{1}{3}R_2 \to R_2$]

Enter: Contents of augmented matrix in cells **A1:C2**

First, modify row 1.

Enter: Formula **=1/2*A1** in cell **A4**

Copy: Contents of cell **A4** to cells **B4:C4**

Second, and this is optional, modify row 2.

Enter: Formula **=1/3*A2** in cell **A5**

Copy: Contents of cell **A5** to cells **B5:C5**

	A	B	C
1	2	-1	4
2	-6	3	-12
3			
4	1	-0.5	2
5	=1/3*A2		

[Perform row operation $2R_1 + R_2 \to R_2$]

First, recopy row 1.

Enter: Formula **=A4** in cell **A7**

Copy: Contents of cell **A7** to cells **B7:C7**

Then, modify row 2.

Enter: Formula **=2*A4+A5** in cell **A8**

Copy: Contents of cell **A8** to cells **B8:C8**

	A	B	C
4	1	-0.5	2
5	-2	1	-4
6			
7	1	-0.5	2
8	=2*A4+A5		

Even using Excel, one must recognize that there are an infinite number of solutions to this system and use the reduced augmented matrix to describe these algebraically.

	A	B	C
7	1	-0.5	2
8	0	0	0

Section 4-3 Gauss-Jordan Elimination

Example 3: Solving a System Using Gauss-Jordan Elimination

[Perform row operation $\frac{1}{2}R_1 \to R_1$]

Enter: Contents of augmented matrix in cells **A1:D3**
First, modify row 1.
Enter: Formula **=1/2*A1** in cell **A5**
Copy: Contents of cell **A5** to cells **B5:D5**
Then, recopy rows 2 and 3.
Enter: Formula **=A2** in cell **A6**
Copy: Contents of cell **A6** to cells **B6:D7**
Remark: Note how the relative cell references work.

	A	B	C	D
1	2	-4	1	-4
2	4	-8	7	2
3	-2	4	-3	5
4				
5	1	-2	0.5	-2
6	4	-8	7	2
7	-2	4	-3	5

[Perform row operations $(-4)R_1 + R_2 \to R_2$ and $2R_1 + R_3 \to R_3$]

First, recopy row 1.
Enter: Formula **=A5** in cell **A9**
Copy: Contents of cell **A9** to cells **B9:D9**
Then, modify rows 2 and 3.
Enter: Formula **=(-4)*A5+A6** in cell **A10**
Copy: Contents of cell **A10** to cells **B10:D10**
Enter: Formula **=2*A5+A7** in cell **A11**
Copy: Contents of cell **A11** to cells **B11:D11**

	A	B	C	D
5	1	-2	0.5	-2
6	4	-8	7	2
7	-2	4	-3	5
8				
9	1	-2	0.5	-2
10	0	0	5	10
11	=2*A5+A7			

[Perform row operation $0.2R_2 \to R_2$]

First, recopy row 1.
Enter: Formula **=A9** in cell **A13**
Copy: Contents of cell **A13** to cells **B13:D13**
Then, modify row 2.
Enter: Formula **=(1/5)*A10** in cell **A14**
Copy: Contents of cell **A14** to cells **B14:D14**
And recopy row 3.
Enter: Formula **=A11** in cell **A15**
Copy: Contents of cell **A15** to cells **B15:D15**

	A	B	C	D
9	1	-2	0.5	-2
10	0	0	5	10
11	0	0	-2	1
12				
13	1	-2	0.5	-2
14	=(1/5)*A10			

[Perform row operations $0.5R_2 + R_1 \to R_1$ and $2R_2 + R_3 \to R_3$]

First, modify row 1.
Enter: Formula **=-0.5*A14+A13** in cell **A17**
Copy: Contents of cell **A17** to cells **B17:D17**
Then, recopy row 2.
Enter: Formula **=A14** in cell **A18**
Copy: Contents of cell **A18** to cells **B18:D18**
And modify row 3.
Enter: Formula **=2*A14+A15** in cell **A19**
Copy: Contents of cell **A19** to cells **B19:D19**
Since the last row produces a contradiction, Gauss-Jordan elimination is stopped.

	A	B	C	D
13	1	-2	0.5	-2
14	0	0	1	2
15	0	0	-2	1
16				
17	1	-2	0	-3
18	0	0	1	2
19	0	0	0	5

Section 4-4 Matrices: Basic Operations

Example 1: Matrix Addition

Enter: Contents of each matrix in cells **A1:C3**
and cells **A4:C5**
Enter: Formula **=A1+A4** in cell **A7**

	A	B	C
1	2	-3	0
2	1	2	-5
3			
4	3	1	2
5	-3	2	5
6			
7	=A1+A4		

Copy: Contents of cell **A7** to cells **A7:C8**

Remark: To copy the contents from **A7** to
A7:C8, drag the contents of cell **A7** to row 7
(**A7:C7**) first. Then, copy the entire row
down to row 8.

	A	B	C
1	2	-3	0
2	1	2	-5
3			
4	3	1	2
5	-3	2	5
6			
7	5	-2	2
8	-2	4	0

Example 2: Matrix Subtraction

Enter: Contents of each matrix in cells
A1:B2 and cells **A4:B5**
Enter: Formula **=A1-A4** in cell **A7**

	A	B
1	3	-2
2	5	0
3		
4	-2	2
5	3	4
6		
7	=A1-A4	

Copy: Contents of cell **A7** to cells **A7:B8**

	A	B
1	3	-2
2	5	0
3		
4	-2	2
5	3	4
6		
7	5	-4
8	2	-4

Example 4: Multiplication of a Matrix by a Number

Enter: Contents of the matrix in cells **A1:C3**
Enter: Formula **=2*A1** in cell **A5**

	A	B	C
1	3	-1	0
2	-2	1	3
3	0	-1	-2
4			
5	=2*A1		

Copy: Contents of cell **A5** to cells **A5:C7**

Remember: To copy the contents from **A5** to **A5:C7**, drag the contents of cell **A5** to row 5 (**A5:C5**) first. Then, copy the entire row down to row 7.

	A	B	C
1	3	-1	0
2	-2	1	3
3	0	-1	-2
4			
5	-6	2	0
6	4	-2	-6
7	0	2	4

Example 5: Sales Commissions

Enter: Contents of the matrices in cells **B3:C4** and **D3:E4**
Enter: Formula **=B3+D3** in cell **B6**

	A	B	C	D	E
1		August Sales		September Sales	
2		Compact	Luxury	Compact	Luxury
3	Smith	$54,000	$88,000	$228,000	$368,000
4	Jones	$126,000	$0	$304,000	$322,000
5		Combined Sales			
6	Smith	=B3+D3			
7	Jones				

Copy: Contents of cell **B6** to cells **B6:C7**

	A	B	C	D	E
1		August Sales		September Sales	
2		Compact	Luxury	Compact	Luxury
3	Smith	$54,000	$88,000	$228,000	$368,000
4	Jones	$126,000	$0	$304,000	$322,000
5		Combined Sales			
6	Smith	$282,000	$456,000		
7	Jones	$430,000	$322,000		

Enter: Formula **=D3-B3** in cell **D6**
Copy: Contents of cell **D6** to cells **D6:E7**

	A	B	C	D	E
1		August Sales		September Sales	
2		Compact	Luxury	Compact	Luxury
3	Smith	$54,000	$88,000	$228,000	$368,000
4	Jones	$126,000	$0	$304,000	$322,000
5		Combined Sales		Sales Increase	
6	Smith	$282,000	$456,000	=D3-B3	
7	Jones	$430,000	$322,000		

Enter: Formula **=0.05*D3** in cell **F3**
Copy: Contents of cell **F3** to cells
F3:G4

	A	B	C	D	E	F	G
1		August Sales		September Sales		September Commissions	
2		Compact	Luxury	Compact	Luxury	Compact	Luxury
3	Smith	$54,000	$88,000	$228,000	$368,000	=0.05*D3	
4	Jones	$126,000	$0	$304,000	$322,000		
5		Combined Sales		Sales Increase			
6	Smith	$282,000	$456,000	$174,000	$280,000		
7	Jones	$430,000	$322,000	$178,000	$322,000		

Example 8: Matrix Multiplication

Enter: Contents of the matrices in cells **A1:B3** and
A5:D6
Select: Cells **A8:D10** (since the resulting matrix
product will be 3 rows by 4 columns)

Select: Cells **A8:D10**. [*Note*: These cells will
receive the matrix product.]

	A	B	C	D
1	2	1		
2	1	0		
3	-1	2		
4				
5	1	-1	0	1
6	2	1	2	0
7				
8				
9				
10				

Matrix multiplication is a built-in Excel function.
To access it, on the **Formulas** tab, select **Math &
Trig.**

Select: **MMULT** from the **Math & Trig** formulas.

Select: **Array1**: Cells **A1:B3**
Note: Use the mouse to select the cells in which
this matrix is located rather than typing **A1:B3**.

Select: **Array2**: Cells **A5:D6**

Do NOT click OK. Instead, use **CRTL-SHIFT-
ENTER** (press **ENTER** while holding **CRTL** and
SHIFT down) rather than just ENTER to input an
array formula.

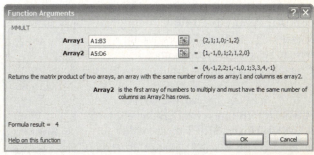

The resulting matrix will appear in cells **A8:D10**.

Alternative method:
Enter: The *array* formula
=MMULT(A1:B3,A5:D6) in selected region (not in just a single cell).
*Use CRTL-SHIFT-ENTER rather than ENTER to input an array formula.

	A	B	C	D
1	2	1		
2	1	0		
3	-1	2		
4				
5	1	-1	0	1
6	2	1	2	0
7				
8	4	-1	2	2
9	1	-1	0	1
10	3	3	4	-1

Example 9: Labor Costs

Enter: Contents of the matrices in cells **B3:C4** and **E3:F4**

Select: Cells **B7:C8** (since the resulting matrix product will be 2 by 2)

	A	B	C	D	E	F
1		Labor-hours per ski			Hourly wages	
2		Assembly	Finishing		California	Maryland
3	Trick ski	5	1.5	Assembly	$12	$13
4	Slalom ski	3	1	Finishing	$7	$8
5		Labor costs per ski				
6		California	Maryland			
7	Trick ski					
8	Slalom ski					

Matrix multiplication is a built-in Excel function. To access it, on the **Formulas** tab, select **Math & Trig.**

Select: **MMULT** from the **Math & Trig** formulas.

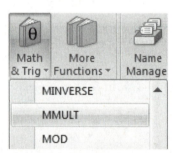

Select: **Array1**: Cells **B3:C4**
Select: **Array2**: Cells **E3:F4**
Remember: To minimize errors, use the mouse to select these arrays rather than typing the cell references.

Do NOT click OK. Instead, use **CRTL-SHIFT-ENTER** (press **ENTER** while holding **CRTL** and **SHIFT** down) rather than just ENTER to input an *array formula.*

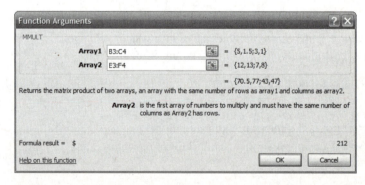

Remember: (1) Highlight the correctly sized cell array before using the MMULT function and (2) use CRTL-SHIFT-ENTER rather than ENTER to input an array formula.

	A	B	C	D	E	F
1		Labor-hours per ski			Hourly wages	
2		Assembly	Finishing		California	Maryland
3	Trick ski	5	1.5	Assembly	$12	$13
4	Slalom ski	3	1	Finishing	$7	$8
5		Labor costs per ski				
6		California	Maryland			
7	Trick ski	$70.50	$77.00			
8	Slalom ski	$43.00	$47.00			

Section 4-5 Inverse of a Square Matrix

Example 2: Finding the Inverse of a Matrix

Enter: Contents of the matrix in cells **A1:C3**
Select: Cells **A5:C7**

	A	B	C
1	1	-1	1
2	0	2	-1
3	2	3	0
4			
5			
6			
7			

Matrix inversion is a built-in Excel function. To access it, on the **Formulas** tab, select **Math & Trig.**

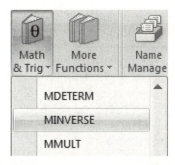

Select: **MINVERSE** from the **Math & Trig** formulas.

Select: **Array1**: Cells **A1:C3**

Remember: To minimize errors, use the mouse to select this array rather than typing the cell references.

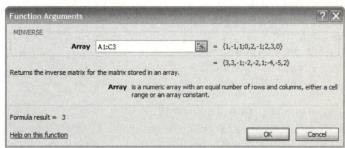

Do NOT click OK. Instead, use **CRTL-SHIFT-ENTER** (press **ENTER** while holding **CRTL** and **SHIFT** down) rather than just ENTER to input an *array formula*. The resulting matrix will appear in cells **A5:C7**.

Alternative method:
Enter: The *array* formula
=MINVERSE(A1:C3) in selected region **A5:C7** (not in just a single cell).
*Use CRTL-SHIFT-ENTER rather than ENTER to input an array formula.

	A	B	C
1	1	-1	1
2	0	2	-1
3	2	3	0
4			
5	3	3	-1
6	-2	-2	1
7	-4	-5	2

Section 4-6 Matrix Equations and Systems of Linear Equations

<u>**Example 2: Using Inverses to Solve Systems of Equations**</u>

Enter: Contents of the coefficient matrix *A* in cells **A1:C3**
Enter: Contents of the matrix *B* in cells **E1:E3**
Select: Cells **A5:C7**

	A	B	C	D	E
1	1	-1	1		1
2	0	2	-1		1
3	2	3	0		1
4					
5					
6					
7					

On the **Formulas** tab, select **Math & Trig.**
Select: **MINVERSE**
In **Function Arguments** window,
Select: **Array1**: Cells **A1:C3**

Do NOT click OK. Instead, use **CRTL-SHIFT-ENTER** (press **ENTER** while holding **CRTL** and **SHIFT** down) rather than just ENTER to input an *array formula*.

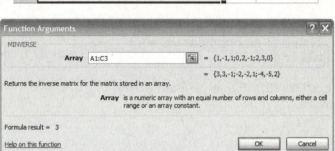

Select: Cells **E5:E7**
On the **Formulas** tab, select **Math & Trig.**
Select: **MMULT**
In **Function Arguments** window,
Select: **Array1**: Cells **A5:C7**
Select: **Array2**: Cells **E1:E3**

Do NOT click OK. Instead, use **CRTL-SHIFT-ENTER** (press **ENTER** while holding **CRTL** and **SHIFT** down) rather than justENTER to input an *array formula*.

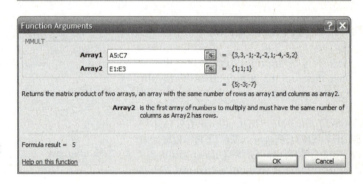

<u>Alternative technique:</u>
Enter: The array formula
=MMULT(MINVERSE(A1:C3),E1:E3) in
selected region **E5:E7** using CRTL-SHIFT-ENTER.

The resulting solution will appear in cells **E5:E7**.

	A	B	C	D	E
1	1	-1	1		1
2	0	2	-1		1
3	2	3	0		1
4					
5	3	3	-1		5
6	-2	-2	1		-3
7	-4	-5	2		-7

Example 4: Investment Analysis

Enter: Data for the problem

	A	B	C	D	E	F	G
1			Clients				
2		1	2	3			
3	Total Investment	$20,000	$50,000	$10,000		1	1
4	Annual Return	$2,400	$7,500	$1,300		0.1	0.2
5	Amount Invested in A						
6	Amount Invested in B						

Select: Cells **B5:B6**
Enter: The array formula
=MMULT(MINVERSE($F$3:$G$4), B3:B4) in selected region.   Use CRTL-SHIFT-ENTER to input an array formula.

	A	B	C	D	E	F	G
1			Clients				
2		1	2	3			
3	Total Investment	$20,000	$50,000	$10,000		1	1
4	Annual Return	$2,400	$7,500	$1,300		0.1	0.2
5	Amount Invested in A	=MMULT(MINVERSE($F$3:$G$4),B3:B4)					
6	Amount Invested in B						

Copy: Contents of cells **B5:B6** to cells **C5:D6**

(Select cells **B5:B6** and drag bottom right corner two cells to right)

	A	B	C	D
1			Clients	
2		1	2	3
3	Total Investment	$20,000	$50,000	$10,000
4	Annual Return	$2,400	$7,500	$1,300
5	Amount Invested in A	$ 16,000	$ 25,000	$ 7,000
6	Amount Invested in B	$ 4,000	$ 25,000	$ 3,000

Section 4-7 Leontief Input-Output Analysis

Example 1: Input-Output Analysis

Enter: The technology matrix T and the final demand data for the problem

	A	B	C	D	E	F
1		Technology Matrix T				Final Demand
2		A	E	M		
3	A	0.2	0	0.1		20
4	E	0.4	0.2	0.1		10
5	M	0	0.4	0.3		30

Enter: Matrix *I-T* in cell **B8:D10**.
Select: Cells **F8:F10**
Enter: The array formula
=MMULT(MINVERSE(B8:D10), F3:F5) in selected region.   Use CRTL-SHIFT-ENTER to input this array formula.

	A	B	C	D	E	F	G	H
1		Technology Matrix T				Final Demand		
2		A	E	M				
3	A	0.2	0	0.1		20		
4	E	0.4	0.2	0.1		10		
5	M	0	0.4	0.3		30		
6								
7	I-T					Output		
8		0.8	0	-0.1		=MMULT(MINVERSE(B8:D10),F3:F5)		
9		-0.4	0.8	-0.1				
10		0	-0.4	0.7				

The output matrix is found in cells
F8:F10.

	A	B	C	D	E	F
1		Technology Matrix T				Final Demand
2		A	E	M		
3	A	0.2	0	0.1		20
4	E	0.4	0.2	0.1		10
5	M	0	0.4	0.3		30
6						
7		I-T				Output
8		0.8	0	-0.1		33
9		-0.4	0.8	-0.1		37
10		0	-0.4	0.7		64

Chapter 5

Section 5-3 Linear Programming in Two Dimensions: A Geometric Approach

Example 2: Solving a Linear Programming Problem

Enter: The corner points of the feasible region *S*.

Enter: Formula **=3*A3+B3** in cell **D3**

Remark: To determine the corner points of the feasible region, use the matrix techniques discussed in Chapter 4.

Copy: Contents of cell **D3** to cells **D4:D5**

	A	B	C	D
1	Corner Point			
2	x	y		z=3x+y
3	3	6		=3*A3+B3
4	2	16		
5	8	4		

	A	B	C	D
1	Corner Point			
2	x	y		z=3x+y
3	3	6		15
4	2	16		22
5	8	4		28

Select: Cell **D7**

On the **Formulas** tab and the **AutoSum** menu, select **Max.**

Using the mouse, select cells **D3:D5.**

The formula **=MAX(D3:D5)** should appear in cell **D7**.

Σ AutoSum Recently Used Financial

Σ Sum
Average
Count Numbers
Max
Min
More Functions...

Select: Cell **D9**

On the **Formulas** tab and the **AutoSum** menu, select **Min.**

Using the mouse, select cells **D3:D5.**

The formula **=MIN(D3:D5)** should appear in cell **D9**.

Alternatively,
Enter: Formula **=MAX(D3:D5)** in cell **D7**
Enter: Formula **=MIN(D3:D5)** in cell **D9**

	A	B	C	D	E
1	Corner Point				
2	x	y		z=3x+y	
3	3	6		15	
4	2	16		22	
5	8	4		28	
6	Max				
7				28	
8	Min				
9				=MIN(D3:D5)	

To find the corner point at which the
minimum and/or maximum occur, use the
LOOKUP function [obviously more useful
for long lists of corner points].

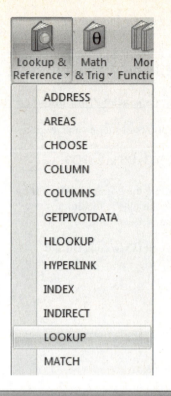

Select: Cell **A7**
On the **Formulas** tab and the **Lookup &
Reference** menu, select **LOOKUP.**

Enter: Formula
=LOOKUP($D7,$D$3:$D$5,A$3:A$5) in
cell **A7** [finds the *x*-value in column **A** that
corresponds to the minimum value in column
D; note the use of anchors ($) going forward]

Copy: Contents of cell **A7** to cell **B7**

In the **Select Arguments** window, select
**Arguments: lookup_value, lookup_vector,
result_vector**
Click: **OK**

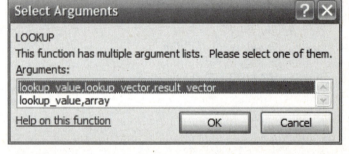

In the **Function Arguments** window,
Enter: Lookup_value: **$D7**
Enter: Lookup_vector: **$D$3:$D$5**
Enter: Result_vector: **A$3:A$5**

Remark: Note the use of anchors ($'s). The
formula in cell **A7** will soon be copied
elsewhere. These anchors will fix either a
column or both a column and a row
(depending on if they appear in front of a
column letter or a row number).

Click: **OK**

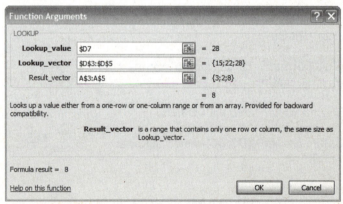

Copy: Contents of cell **A7** to cells **B7, A9** and **B9**

	A	B	C	D	E
1	Corner Point				
2	x	y		z=3x+y	
3	3	6		15	
4	2	16		22	
5	8	4		28	
6	Max				
7	=LOOKUP($D7,$D$3:$D$5,A$3:A$5)				

Results appear in rows 7 and 9.

The maximum value is located at (8,4) and has a value of 28.

The minimum value is located at (3,6) and has a value of 15.

	A	B	C	D
1	Corner Point			
2	x	y		z=3x+y
3	3	6		15
4	2	16		22
5	8	4		28
6	Max			
7	8	4		28
8	Min			
9	3	6		15

Chapter 6

Section 6-2 Simplex Method: Maximization with Problem Constraints of the Form ≤

Example 1: Using the Simplex Method

Enter: The initial simplex tableau and identify the first pivot element.

	A	B	C	D	E	F	G
1		x_1	x_2	s_1	s_2	P	
2	s_1	4	1	1	0	0	28
3	s_2	2	3	0	1	0	24
4	P	-10	-5	0	0	1	0

Perform row operation: $\frac{1}{4}R_1 \to R_1$

First, modify row 1.
Enter: **=1/4*B2** in cell **B6**
Copy: Contents of cell **B6** to cells **C6:G6**

	A	B	C	D	E	F	G
1		x_1	x_2	s_1	s_2	P	
2	s_1	4	1	1	0	0	28
3	s_2	2	3	0	1	0	24
4	P	-10	-5	0	0	1	0
5							
6	s_1	=1/4*B2					
7	s_2						
8	P						

Then, recopy rows 2 and 3.
Enter: **=B3** in cell **B7**
Copy: Contents of cell **B7** to cells **B7:G8**

	A	B	C	D	E	F	G
1		x_1	x_2	s_1	s_2	P	
2	s_1	4	1	1	0	0	28
3	s_2	2	3	0	1	0	24
4	P	-10	-5	0	0	1	0
5							
6	s_1	1	0.25	0.25	0	0	7
7	s_2	2	3	0	1	0	24
8	P	-10	-5	0	0	1	0

Perform row operations:
$(-2)R_1 + R_2 \to R_2$ and $10R_1 + R_3 \to R_3$

First, recopy row 1.
Enter: **=B6** in cell **B10**
Copy: Contents of cell **B10** to cells **C10:G10**
Then, modify rows 2 and 3.
Enter: **=(-2)*B6+B7** in cell **B11**
Copy: Contents of cell **B11** to cells **C11:G11**
Enter: **=10*B6+B8** in cell **B12**
Copy: Contents of cell **B12** to cells **C12:G12**

	A	B	C	D	E	F	G
6	s_1	1	0.25	0.25	0	0	7
7	s_2	2	3	0	1	0	24
8	P	-10	-5	0	0	1	0
9							
10	x_1	1	0.25	0.25	0	0	7
11	s_2	=-2*B6+B7					
12	P						

	A	B	C	D	E	F	G
6	s_1	1	0.25	0.25	0	0	7
7	s_2	2	3	0	1	0	24
8	P	-10	-5	0	0	1	0
9							
10	x_1	1	0.25	0.25	0	0	7
11	s_2	0	2.5	-0.5	1	0	10
12	P	=10*B6+B8					

To determine next pivot element, look above the negative element in the bottom row.
Enter: Formula **=G10/C10** in cell **I10**
Copy: Contents of cell **I10** to cell **I11**

	A	B	C	D	E	F	G	H	I
10	x_1	1	0.25	0.25	0	0	7		=G10/C10
11	s_2	0	2.5	-0.5	1	0	10		
12	P	0	-2.5	2.5	0	1	70		

	A	B	C	D	E	F	G	H	I
10	x_1	1	0.25	0.25	0	0	7		28
11	s_2	0	2.5	-0.5	1	0	10		4
12	P	0	-2.5	2.5	0	1	70		

Perform row operation: $\frac{1}{2.5}R_2 \rightarrow R_2$

Enter: **=1/2.5*B11** in cell **B15**
Copy: Contents of cell **B15** to cells **C15:G15**
Enter: **=B10** in cell **B14** and **=B12** in cell **B16**
Copy: Contents of cell **B14** to cells **C14:G14** and the contents of cell **B16** to cells **C16:G16**
Perform row operations:
$-0.25R_2 + R_1 \rightarrow R_1$ and $2.5R_2 + R_3 \rightarrow R_3$
Enter: **=-0.25*B15+B14** in cell **B18**
Copy: Contents of cell **B18** to cells **C18:G18**
Enter: **=B15** in cell **B19** and **=2.5*B15+B16** in cell **B20**
Copy: Contents of cell **B19** to cells **C19:G19** and the contents of cell **B20** to cells **C20:G20**
Since all the indicators in the last row are nonnegative, the optimal solution has been found.

	A	B	C	D	E	F	G
10	x_1	1	0.25	0.25	0	0	7
11	s_2	0	2.5	-0.5	1	0	10
12	P	0	-2.5	2.5	0	1	70
13							
14	x_1	1	0.25	0.25	0	0	7
15	s_2	0	1	-0.2	0.4	0	4
16	P	0	-2.5	2.5	0	1	70

	A	B	C	D	E	F	G
14	x_1	1	0.25	0.25	0	0	7
15	s_2	0	1	-0.2	0.4	0	4
16	P	0	-2.5	2.5	0	1	70
17							
18	x_1	1	0	0.3	-0.1	0	6
19	s_2	0	1	-0.2	0.4	0	4
20	P	=2.5*B15+B16					

	A	B	C	D	E	F	G
18	x_1	1	0	0.3	-0.1	0	6
19	s_2	0	1	-0.2	0.4	0	4
20	P	0	0	2	1	1	80

Example 1: Using the Simplex Method (alternative method using Excel's Solver)

Enter: Any value (e.g. **0**) in cell **A2**
Enter: Any value (e.g. **0**) in cell **B2**
Enter: Formula **=10*A2+5*B2** in cell **C2** (representing the value of *P*)

Enter: Formula **=4*A2+B2** in cell **A5** (representing the left hand side of inequality 1)
Enter: Value **28** in cell **B5** (representing the right hand side of inequality 1)
Enter: Formula **=2*A2+3*B2** in cell **A6** (representing the left hand side of inequality 2)
Enter: Value **24** in cell **B6**
The **Solver** Add-in for Excel can help analyze systems of equations. *If this Add-in has not been installed before*, you will need to install it.

Select: Large, round **Office Button**
Select: **Excel Options** button at bottom

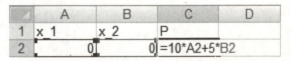

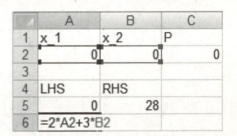

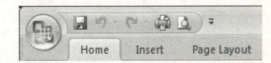

In the **Excel Options** window,

Select: **Add-Ins** on the left menu
Select: **Solver Add-in**
Click: **Go...**

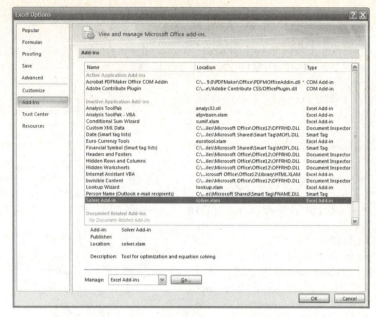

In the **Add-Ins** window,

Select: **Solver Add-in**
Click: **OK**

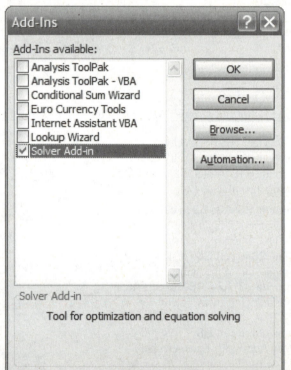

In the **Microsoft Office Excel** window,

Select: **Yes**

Select: **Data** tab

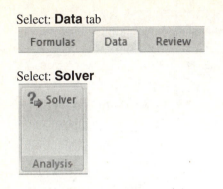

Select: **Solver**

Enter: **Select Target Cell: C2**
Select: **Equal To: Max**
Enter: **By Changing Cells: A2, B2**
Click: Subject to the Constraints: **Add**

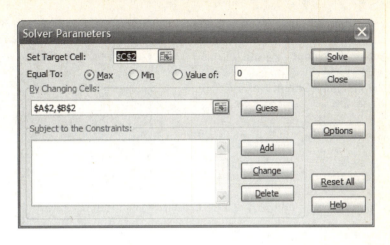

Enter: Cell Reference: **A5**
Select: **<=**
Enter: Constraint: **B5**
Click: **Add**

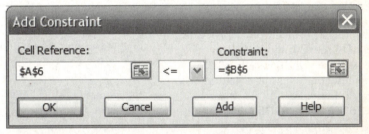

Enter: Cell Reference: **A6**
Select: **<=**
Enter: Constraint: **B6**
Click: **Add**

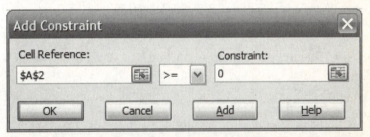

Enter: Cell Reference: **A2**
Select: **>=**
Enter: Constraint: **0**
Click: **Add**

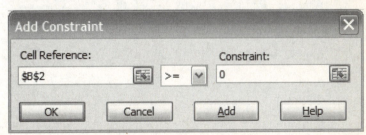

Enter: Cell Reference: **B2**
Select: **>=**
Enter: Constraint: **0**
Click: **OK**

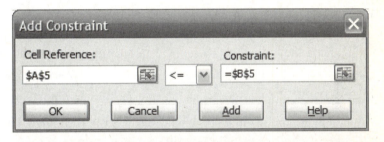

Click: **Solve**

When the **Solver Results** window appears,
Select: **Keep Solver Solution**
Click: **OK**

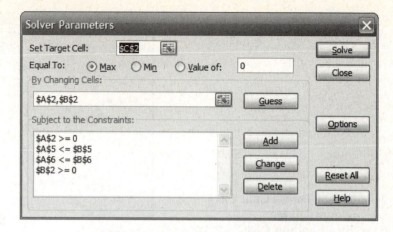

The optimal solution appears in cells **A2**, **B2**, and the optimal value appears in cell **C2**.

	A	B	C
1	x_1	x_2	P
2	6	4	80
3			
4	LHS	RHS	
5	28	28	
6	24	24	

Example 3: Agriculture

Enter: Data from the problem in cells **B2:D4, E2:E4,** and **B5:D5**
Enter: The value **0** in cells **B6:D6** (representing a feasible solution)
Enter: Formula
=B5*B6+C5*C6+D5*D6 in cell **E5** (representing the value of *P*)

	A	B	C	D	E	F	G
1	Resources	Crop A	Crop B	Crop C	Available	Used	
2	Acres	1	1	1	100		
3	Seed ($)	40	20	30	3200		
4	Workdays	1	2	1	160		
5	Profit Per Acre	100	300	200	=B5*B6+C5*C6+D5*D6		
6	Acres to plant	0	0	0		Profit	

Enter: Formula **=B2*$B$6+C2*$C$6+D2*$D$6** in cell **F2** (representing the left hand side of inequality 1)
Copy: The contents of cell **F2** to cells **F3:F4** (representing inequalities 2 and 3)

	A	B	C	D	E	F	G	H
1	Resources	Crop A	Crop B	Crop C	Available	Used		
2	Acres	1	1	1	100	=B2*$B$6+C2*$C$6+D2*$D$6		
3	Seed ($)	40	20	30	3200			
4	Workdays	1	2	1	160			
5	Profit Per Acre	100	300	200	0	<-Total		
6	Acres to plant	0	0	0		Profit		

Select: **Data** tab

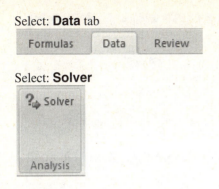

Select: **Solver**

Enter: **Select Target Cell: E5**
Select: **Equal To: Max**
Enter: **By Changing Cells: B6:D6**
Click: Subject to the Constraints: **Add**

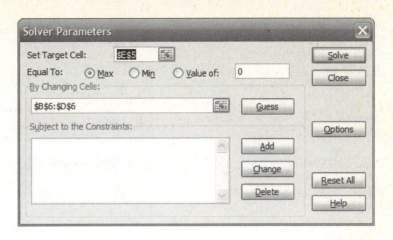

Enter: Cell Reference: **F2**
Select: **<=**
Enter: Constraint: **E2**
Click: **Add**

Enter: Cell Reference: **F3**
Select: **<=**
Enter: Constraint: **E3**
Click: **Add**

Enter: Cell Reference: **F4**
Select: **<=**
Enter: Constraint: **E4**
Click: **Add**

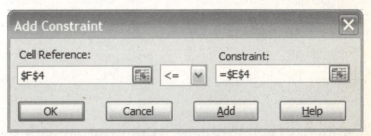

Enter: Cell Reference: **B6**
Select: **>=**
Enter: Constraint: **0**
Click: **Add**

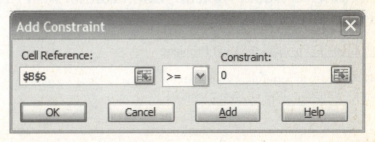

Enter: Cell Reference: **C6**
Select: **>=**
Enter: Constraint: **0**
Click: **Add**

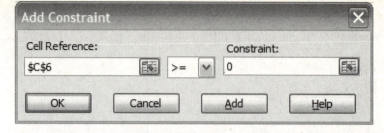

Enter: Cell Reference: **D6**
Select: **>=**
Enter: Constraint: **0**
Click: **OK**

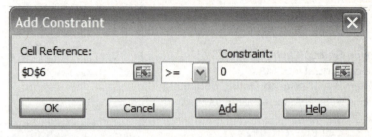

Click: **Solve**

When the **Solver Results** window
appears,
Select: **Keep Solver Solution**
Click: **OK**

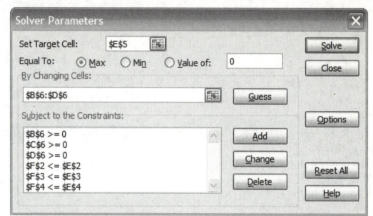

The optimal solution appears in cells **B6**,
C6, and **D6** and the optimal value
appears in cell **E5**. Also, the amount of
each resource used appears in column **F**.

	A	B	C	D	E	F
1	Resources	Crop A	Crop B	Crop C	Available	Used
2	Acres	1	1	1	100	100
3	Seed ($)	40	20	30	3200	2400
4	Workdays	1	2	1	160	160
5	Profit Per Acre	100	300	200	26000	<-Total
6	Acres to plant	0	60	40		Profit

Section 6-3 Dual Problem: Minimization with Problem Constraints of the Form ≥

Example 1: Forming the Dual Problem

Though easy to perform by hand, Excel
easily computes the transpose of a matrix.

	A	B	C	D
1	2	1	5	20
2	4	1	1	30
3	40	12	40	1

Matrix transposition is a built-in Excel function. To access it, on the **Formulas** tab, select **Lookup & Reference.**

Select: **TRANSPOSE** from the **Lookup & Reference** formulas.

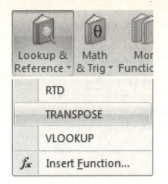

Select: **Array**: Cells **A1:D3**

Remember: To minimize errors, use the mouse to select this array rather than typing the cell references.

Do NOT click OK. Instead, use **CRTL-SHIFT-ENTER** (all three keys at the same time) rather than ENTER to input an *array formula*.

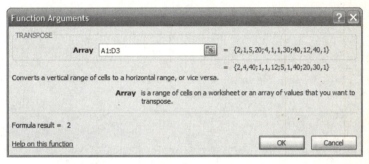

The resulting matrix will appear in cells **A5:C8**.

Alternative method:
Enter: The *array* formula
=TRANSPOSE(A1:C3) in selected region **A5:C8** (not in just a single cell).
*Use CRTL-SHIFT-ENTER rather than ENTER to input an array formula.

	A	B	C	D
1	2	1	5	20
2	4	1	1	30
3	40	12	40	1
4				
5	2	4	40	
6	1	1	12	
7	5	1	40	
8	20	30	1	

Example 4: Transportation Problem

Enter: Data from problem in cells **B5:C6, D5:D6**, and **B7:C7**
Enter: Value **0** in cells **G5:H6**
Enter: Formula **=G5+H5** in cell **I5**
Copy: Formula in cell **I5** to cell **I6**
Enter: Formula **=G5+G6** in cell **G7**
Copy: Formula in cell **G7** to cell **H7**

	A	B	C	D	E	F	G	H	I
1		Data					Shipping Schedule		
2		Distribution					Distribution		
3		Outlet		Assembly			Outlet		
4		I	II	Capacity			I	II	Total
5	Plant A	$6	$5	700		Plant A	0	0	=G5+H5
6	Plant B	$4	$8	900		Plant B	0	0	
7	Minimum	500	1000			Total	0	0	

Enter: Formula **=B5*G5+C5*H5+B6*G6+C6*H6** in cell **I8** (representing the value of *C*)

	A	B	C	D	E	F	G	H	I	J	K
1		Data					Shipping Schedule				
2		Distribution					Distribution				
3		Outlet		Assembly			Outlet				
4		I	II	Capacity			I	II	Total		
5	Plant A	$6	$5	700		Plant A	0	0	0		
6	Plant B	$4	$8	900		Plant B	0	0	0		
7	Minimum	500	1000			Total	0	0			
8	Required					Total Cost			=B5*G5+C5*H5+B6*G6+C6*H6		

Enter: Formula **=G5+H5** in cell **A12**
Enter: Formula **=D5** in cell **B12**
Enter: Formula **=G6+H6** in cell **A13**
Enter: Formula **=D6** in cell **B13**
Enter: Formula **=G5+G6** in cell **A14**
Enter: Formula **=B7** in cell **B14**
Enter: Formula **=H5+H6** in cell **A15**
Enter: Formula **=C7** in cell **B15**

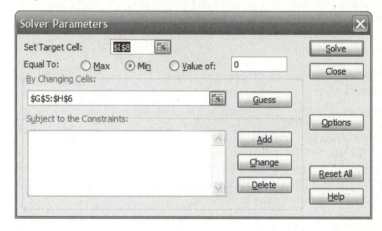

	A	B	C	D
1		Data		
2		Distribution		
3		Outlet		Assembly
4		I	II	Capacity
5	Plant A	$6	$5	700
6	Plant B	$4	$8	900
7	Minimum	500	1000	
8	Required			
9				
10	Constraints			
11	LHS	RHS		
12	0	700		
13	0	900		
14	0	500		
15	0	=C7		

Select: **Data** tab

Select: **Solver**

Enter: **Select Target Cell: I8**
Select: **Equal To: Min**
Enter: **By Changing Cells: G5:H6**
Click: Subject to the Constraints: **Add**

Enter: Cell Reference: **A12**
Select: **<=**
Enter: Constraint: **B12**
Click: **Add**

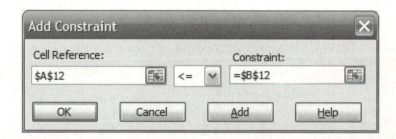

Repeat above step adding constraints:
A13 <= B13
A14 >= B14
A15 >= B15
G5 >= 0
H5 >= 0
G6 >= 0
H6 >= 0
Click: **OK**

Click: **Solve**

When the **Solver Results** window appears,
Select: **Keep Solver Solution**
Click: **OK**

The optimal solution appears in cells **G5**, **G6**, **H5**, and **H6** and the optimal value appears in cell **I8**.

	A	B	C	D	E	F	G	H	I
1		Data					Shipping Schedule		
2		Distribution					Distribution		
3		Outlet		Assembly			Outlet		
4		I	II	Capacity			I	II	Total
5	Plant A	$6	$5	700		Plant A	0	700	700
6	Plant B	$4	$8	900		Plant B	500	300	800
7	Minimum	500	1000			Total	500	1000	
8	Required					Total Cost			$7,900
9									
10	Constraints								
11	LHS	RHS							
12	700	700							
13	800	900							
14	500	500							
15	1000	1000							

Section 6-4 Maximization and Minimization with Mixed Problem Constraints

Example 2: Using the Big *M* Method

Enter: The preliminary simplex tableau for the modified problem.
Enter: Any relatively large number (compared to numbers in the tableau) in cell **L1**
Enter: Formula **=$L$2** (absolute reference) in cells **F5** and **G5**

	A	B	C	D	E	F	G	H	I	J	K	L
1		x_1	x_2	x_3	s_1	a_1	s_2	a_2	P			1000
2	s_1	1	1	0	1	0	0	0	0	20		
3	a_1	1	0	1	0	1	0	0	0	5		
4	a_2	0	1	1	0	0	-1	1	0	10		
5		-1	1	-3	0	=$L$1						

First, recopy the tableau.
Enter: Formula **=B2** in cell **B7**
Copy: Contents of cell **B7** to cells **B7:J9**
Then, update the bottom row.
Enter: Formula **=-$L$1*B3+B5** in cell **B10**
Copy: Contents of cell **B10** to cells **C10:J10**

	A	B	C	D	E	F	G	H	I	J	K	L
1		x_1	x_2	x_3	s_1	a_1	s_2	a_2	P			1000
2	s_1	1	1	0	1	0	0	0	0	20		
3	a_1	1	0	1	0	1	0	0	0	5		
4	a_2	0	1	1	0	0	-1	1	0	10		
5		-1	1	-3	0	1000	0	1000	1	0		
6												
7	s_1	1	1	0	1	0	0	0	0	20		
8	a_1	1	0	1	0	1	0	0	0	5		
9	a_2	0	1	1	0	0	-1	1	0	10		
10		=-$L$1*B3+B5										

First, recopy the tableau.
Enter: Formula **=B7** in cell **B12**
Copy: Contents of cell **B12** to cells **B12:J14**
Then, update the last row.
Enter: Formula **=-$L$1*B9+B10** in cell **B15**
Copy: Contents of cell **B15** to cells **C15:J15**

	A	B	C	D	E	F	G	H	I	J
7	s_1	1	1	0	1	0	0	0	0	20
8	a_1	1	0	1	0	1	0	0	0	5
9	a_2	0	1	1	0	0	-1	1	0	10
10		-1001	1	-1003	0	0	0	1000	1	-5000
11										
12	s_1	1	1	0	1	0	0	0	0	20
13	a_1	1	0	1	0	1	0	0	0	5
14	a_2	0	1	1	0	0	-1	1	0	10
15		=-$L$1*B9+B10								

Pivot on the second row and third column. Perform the row operations $(-1)R_2 + R_3 \rightarrow R_3$ and

$(2M+3)R_2 + R_4 \rightarrow R_4$.

Enter: Formula **=B12** in cell **B17**
Copy: Contents of cell **B17** to cells **B17:J18**
Enter: Formula **=-B13+B14** in cell **B19**
Copy: Contents of cell **B19** to cells **C19:J19**
Enter: Formula **=(2*$L$1+3)*B13+B15** in cell **B20**
Copy: Contents of cell **B20** to cells **C20:J20**

	A	B	C	D	E	F	G	H	I	J
12	s_1	1	1	0	1	0	0	0	0	20
13	a_1	1	0	1	0	1	0	0	0	5
14	a_2	0	1	1	0	0	-1	1	0	10
15		-1001	-999	-2003	0	0	1000	0	1	-15000
16										
17	s_1	1	1	0	1	0	0	0	0	20
18	x_3	1	0	1	0	1	0	0	0	5
19	a_2	-1	1	0	0	-1	-1	1	0	5
20		=(2*$L$1+3)*B13+B15								

Pivot on the third row and second column. Perform the row operations $(-1)R_3 + R_1 \rightarrow R_1$ and

$(M-1)R_3 + R_4 \rightarrow R_4$.

Enter: Formula **=B12** in cell **B17**
Copy: Contents of cell **B17** to cells **B17:J18**
Enter: Formula **=-B13+B14** in cell **B19**
Copy: Contents of cell **B19** to cells **C19:J19**
Enter: Formula **=(2*$L$1+3)*B13+B15** in cell **B20**
Copy: Contents of cell **B20** to cells **C20:J20**

	A	B	C	D	E	F	G	H	I	J
17	s_1	1	1	0	1	0	0	0	0	20
18	x_3	1	0	1	0	1	0	0	0	5
19	a_2	-1	1	0	0	-1	-1	1	0	5
20		1002	-999	0	0	2003	1000	0	1	-4985
21										
22	s_1	2	0	0	1	1	1	-1	0	15
23	x_3	1	0	1	0	1	0	0	0	5
24	x_2	-1	1	0	0	-1	-1	1	0	5
25		3	0	0	0	1004	1	999	1	10

Since the bottom row has no negative indicators, the optimal solution can be read from the tableau.

Example 3: Using the Big *M* Method

Enter: Any value (e.g. **0**) in cell **A2**
Enter: Any value (e.g. **0**) in cell **B2**
Enter: Formula **=3*A2+5*B2** in cell **C2**
(representing the value of *P*)

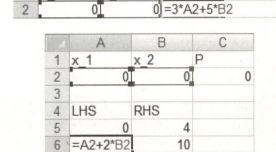

Enter: Formula **=2*A2+B2** in cell **A5**
(representing the left hand side of
inequality 1)
Enter: Value **4** in cell **B5**
Enter: Formula **=A2+2*B2** in cell **A6**
(representing the left hand side of
inequality 2)
Enter: Value **10** in cell **B6**

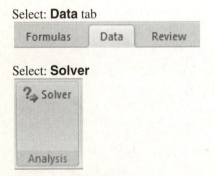

Select: **Data** tab

Select: **Solver**

Enter: **Select Target Cell: C2**
Select: **Equal To: Max**
Enter: **By Changing Cells: A2,B2**
Click: Subject to the Constraints: **Add**

Enter: Cell Reference: **A5**
Select: **<=**
Enter: Constraint: **B5**
Click: **Add**

Repeat above step adding constraints:
A6 >= B6
A2 >= 0
B2 >= 0
Click: **OK**

Click: **Solve**

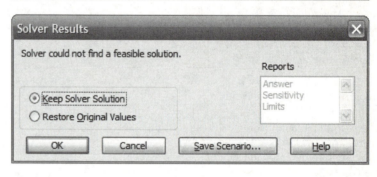

When the **Solver Results** window
appears,
Select: **Keep Solver Solution**
Click: **OK**

Excel does not find a feasible solution
(and rightfully so since no feasible
solution exists).

Chapter 7

Section 7-1 Logic

Example 1: Compound Propositions

Enter: Value **FALSE** in cell **A2** (since proposition *p* is false)
Enter: Value **TRUE** in cell **B2** (since proposition *q* is true)

Excel has several built-in logical functions. To access them, on the **Formulas** tab, select **Logical.**

Select: Cell **A5**
Select: **NOT** from the **Logical** formulas.

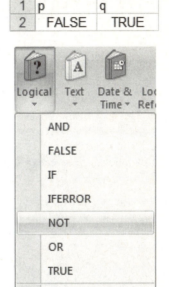

In **Function Arguments** window,
Enter: **Logical: A2**
Click: **OK**

Remember: Use the mouse and select cell **A2** rather than typing **A2.**

Repeat this process in cell **B5** by entering **Logical: B2** in the **Function Arguments** window.

Alternatively:
Enter: Formula **=NOT(A2)** in cell **A5**
Enter: Formula **=NOT(B2)** in cell **B5**

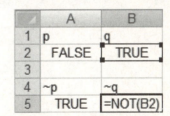

Select: Cell **C5**
Select: **OR** from the **Logical**
formulas.

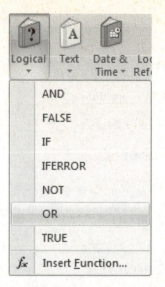

In **Function Arguments** window,
Enter: **Logical1: A2**
Enter: **Logical2: B2**
Click: **OK**

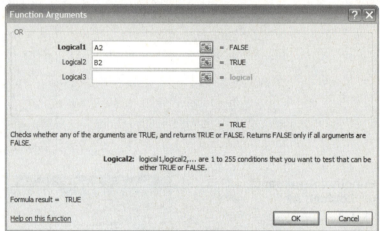

Alternatively:
Enter: Formula **=OR(A2,B2)** in cell
C5

	A	B	C	D	E
1	p	q			
2	FALSE	TRUE			
3					
4	~p	~q	p v q	p ^ q	p -> q
5	TRUE	FALSE	=OR(A2,B2)		

Repeat the previous process in cell **D5**
using the **AND** function from the
formulas menu.

Alternatively:
Enter: Formula **=AND(A2,B2)** in cell
D5

	A	B	C	D	E
1	p	q			
2	FALSE	TRUE			
3					
4	~p	~q	p v q	p ^ q	p -> q
5	TRUE	FALSE	TRUE	=AND(A2,B2)	

Select: Cell **E5**
Select: **IF** from the **Logical** formulas.

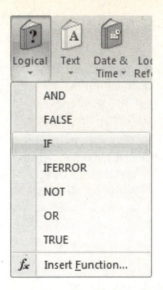

In **Function Arguments** window,
Enter: **Logical test: A2**
Enter: **Value if true: B2**
Enter: **Value if false: TRUE**
Click: **OK**

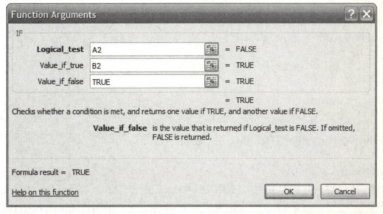

Alternatively:
Enter: Formula **=IF(A2,B2,TRUE)** in
cell **D5** (remember that if *p* is false,
$p \rightarrow q$ is vacuously true)

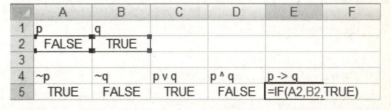

	A	B	C	D	E	F
1	p	q				
2	FALSE	TRUE				
3						
4	~p	~q	p v q	p ^ q	p -> q	
5	TRUE	FALSE	TRUE	FALSE	=IF(A2,B2,TRUE)	

Example 2: Converse and Contrapositive

Enter: Value **TRUE** in cell **A2** (since
proposition *p* is true)
Enter: Value **FALSE** in cell **B2** (since
proposition *q* is false)
Enter: Formula **=IF(A2,B2,TRUE)** in
cell **A5**

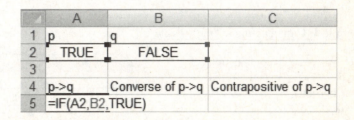

	A	B	C
1	p	q	
2	TRUE	FALSE	
3			
4	p->q	Converse of p->q	Contrapositive of p->q
5	=IF(A2,B2,TRUE)		

Remark: Using the **Logical** menu on the
Formulas tab to enter this last formula
involving the **IF** function is advised.

Enter: Formula **=IF(B2,A2,TRUE)** in cell **B5**

Remark: Again, using the **Logical** menu on the **Formulas** tab to enter this last formula involving the **IF** function is advised.

	A	B	C
1	p	q	
2	TRUE	FALSE	
3			
4	p->q	Converse of p->q	Contrapositive of p->q
5	FALSE	=IF(B2,A2,TRUE)	

Enter: Formula **=IF(NOT(B2),NOT(A2),TRUE)** in cell **C5**

	A	B	C	D
1	p	q		
2	TRUE	FALSE		
3				
4	p->q	Converse of p->q	Contrapositive of p->q	
5	FALSE	TRUE	=IF(NOT(B2),NOT(A2),TRUE)	

Example 3: Constructing Truth Tables

Enter: All possible combinations of variables *p* and *q* in cells **A2:B5**

	A	B	C	D
1	p	q	~p	~p v q
2	TRUE	TRUE		
3	TRUE	FALSE		
4	FALSE	TRUE		
5	FALSE	FALSE		

Enter: Formula **=NOT(A2)** in cell **C2**

Alternatively:
Use the **Logical** menu on the **Formulas** tab to enter this formula involving the **NOT** function.

	A	B	C	D
1	p	q	~p	~p v q
2	TRUE	TRUE	FALSE	
3	TRUE	FALSE	FALSE	
4	FALSE	TRUE	TRUE	
5	FALSE	FALSE	TRUE	

Copy: Contents of cell **C2** to cells **C3:C5**

Enter: Formula **=OR(C2,B2)** in cell **D2**
Copy: Contents of cell **D2** to cells **D3:D5**

Alternatively:
Use the **Logical** menu on the **Formulas** tab to enter this formula involving the **OR** function.

	A	B	C	D	E
1	p	q	~p	~p v q	
2	TRUE	TRUE	FALSE	=OR(C2,B2)	
3	TRUE	FALSE	FALSE		
4	FALSE	TRUE	TRUE		
5	FALSE	FALSE	TRUE		

Example 4: Constructing a Truth Table

Enter: All possible combinations of variables *p* and *q* in cells **A2:B5**
Enter: Formula **=IF(A2,B2,TRUE)** in cell **C2**
Copy: Contents of cell **C2** to cells **C3:C5**

	A	B	C	D	E
1	p	q	p->q	(p->q) ^ p	[(p->q) ^ p]->q
2	TRUE	TRUE	=IF(A2,B2,TRUE)		
3	TRUE	FALSE			
4	FALSE	TRUE			
5	FALSE	FALSE			

Enter: Formula **=AND(C2,A2)** in cell **D2**
Copy: Contents of cell **D2** to cells **D3:D5**

	A	B	C	D	E
1	p	q	p->q	(p->q) ^ p	[(p->q) ^ p]->q
2	TRUE	TRUE	TRUE	=AND(C2,A2)	
3	TRUE	FALSE	FALSE		
4	FALSE	TRUE	TRUE		
5	FALSE	FALSE	TRUE		

Enter: Formula **=IF(D2,B2,TRUE)** in cell **E2**
Copy: Contents of cell **E2** to cells **E3:E5**

	A	B	C	D	E	F
1	p	q	p->q	(p->q) ^ p	[(p->q) ^ p]->q	
2	TRUE	TRUE	TRUE	TRUE	=IF(D2,B2,TRUE)	
3	TRUE	FALSE	FALSE	FALSE		
4	FALSE	TRUE	TRUE	FALSE		
5	FALSE	FALSE	TRUE	FALSE		

Example 6: Verifying a Logical Implication

Enter: All possible combinations of variables *p* and *q* in cells **A2:B5**
Enter: Formula **=IF(A2,B2,TRUE)** in cell **C2**
Copy: Contents of cell **C2** to cells **C3:C5**

	A	B	C	D	E	F
1	p	q	p->q	~q	(p->q) ^ ~q	~p
2	TRUE	TRUE	=IF(A2,B2,TRUE)			
3	TRUE	FALSE				
4	FALSE	TRUE				
5	FALSE	FALSE				

Enter: Formula **=NOT(B2)** in cell **D2**
Copy: Contents of cell **D2** to cells **D3:D5**

	A	B	C	D	E	F
1	p	q	p->q	~q	(p->q) ^ ~q	~p
2	TRUE	TRUE	TRUE	=NOT(B2)		
3	TRUE	FALSE	FALSE			
4	FALSE	TRUE	TRUE			
5	FALSE	FALSE	TRUE			

Enter: Formula **=AND(C2,D2)** in cell **E2**
Copy: Contents of cell **E2** to cells **E3:E5**

	A	B	C	D	E	F
1	p	q	p->q	~q	(p->q) ^ ~q	~p
2	TRUE	TRUE	TRUE	FALSE	=AND(C2,D2)	
3	TRUE	FALSE	FALSE	TRUE		
4	FALSE	TRUE	TRUE	FALSE		
5	FALSE	FALSE	TRUE	TRUE		

Enter: Formula **=NOT(A2)** in cell **F2**
Copy: Contents of cell **F2** to cells **F3:F5**

	A	B	C	D	E	F
1	p	q	p->q	~q	(p->q) ^ ~q	~p
2	TRUE	TRUE	TRUE	FALSE	FALSE	FALSE
3	TRUE	FALSE	FALSE	TRUE	FALSE	FALSE
4	FALSE	TRUE	TRUE	FALSE	FALSE	TRUE
5	FALSE	FALSE	TRUE	TRUE	TRUE	TRUE

Section 7-4 Permutations and Combinations

Example 1: Computing Factorials

Excel has several built-in counting functions. To access them, on the **Formulas** tab, select **Math & Trig**.

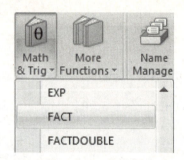

To compute 5!
Select: Cell **A1**
Select: **FACT** from the **Math & Trig** formulas.

In the **Function Arguments** window,
Enter: **Number: 5**
Click: **OK**

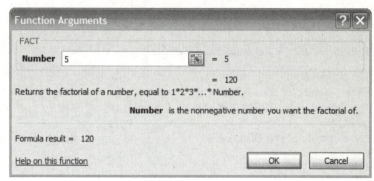

Alternatively:
Enter: Formula **=FACT(5)** in cell **A1**

	A
1	=FACT(5)

To compute $\dfrac{7!}{6!}$:

Enter: Formula **=FACT(7)/FACT(6)** in cell **A1**

	A
1	=FACT(7)/FACT(6)

Example 3: Permutations

Excel has several built-in statistical functions. To access them, on the **Formulas** tab, select **More Functions -> Statistical**.

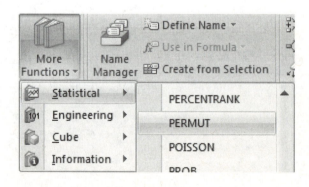

To compute $P_{13,8}$:
Select: Cell **A1**
Select: **PERMUT** from the **Statistical** formulas.

In the **Function Arguments** window,
Enter: **Number: 13**
Enter: **Number chosen: 8**
Click: **OK**

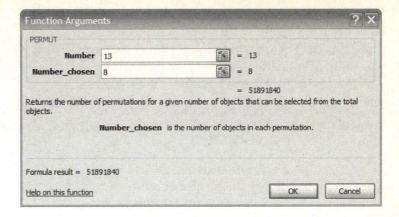

Alternatively:
Enter: Formula **=PERMUT(13,8)** in cell
A1

Example 5: Combinatations

Interestingly, Excel's built-in combination function is *not* found under the statistical functions (as the permutations function is). Instead, it can be found under the **Formulas** tab and **Math & Trig** menu.

To compute $C_{13,8}$:

Select: Cell **A1**
Select: **COMBIN** from the **Math & Trig** formulas.
In the **Function Arguments** window,
Enter: **Number: 13**
Enter: **Number chosen: 8**
Click: **OK**

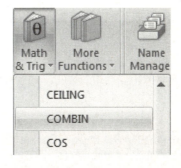

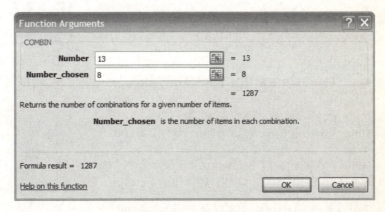

Alternatively:
Enter: Formula **=COMBIN(13,8)** in cell
A1

Chapter 8

Section 8-1 Sample Spaces, Events, and Probability

Example 6: Simulation and Empirical Probabilities

The built-in Excel **RAND** function generates a random number between 0 and 1. To access it, on the **Formulas** tab, select **Math & Trig.**

Multiplying the output of the **RAND** function by 6 has the effect of generating a random number between 0 and 6. Rounding up to 0 digits yields a random integer between 1 and 6.

To simulate 100 rolls of two dice,
Select: Cell **A1**
On the **Formulas** tab, select **ROUNDUP** under the **Math & Trig** menu.

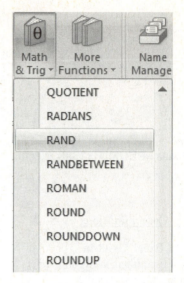

In the Function Arguments window,
Enter: **Number: 6*RAND()**
Enter: **Num digits: 0**
Click: **OK**

Remark: The **()** after the **RAND** function denote the fact that the function takes no arguments.

Alternatively:
Enter: Formula **=ROUNDUP(6*RAND(),0)** in cell **A1**

Copy: Contents of cell **A1** to cell **B1**
Copy: Contents of cells **A1:B1** to cells **A2:B100**

Column **A** represents 100 rolls of one die and column **B** represents 100 rolls of the other.

Remember: To copy these values, simply hold the mouse button over the lower right corner of the cell and drag.

Enter: Formula **=SUM(A1:B1)** in cell **C1**
Copy: Contents of cell **C1** to cells **C2:C100**
Column **C** represents the sum on the dice for each of the 100 rolls.

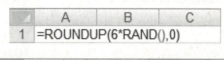

	A	B	C
1	=ROUNDUP(6*RAND(),0)		

	A	B	C	D
1	3	2	=SUM(A1:B1)	
2	2	4		
3	4	6		
4	3	6		
5	4	3		

Now, set up a frequency table:
Enter: Value **2** in cell **E2**
Enter: Formula **=E2+1** in cell **E3**
Copy: Contents of cell **E3** to cells **E4:E12**

	A	B	C	D	E	F
1	2	3	5		Total	Frequency
2	6	2	8		2	
3	6	6	12		3	
4	5	3	8		4	
5	5	2	7		5	
6	5	2	7		6	
7	5	6	11		7	
8	2	4	6		8	
9	2	4	6		9	
10	6	1	7		10	
11	5	6	11		11	
12	3	4	7		12	

Select: Cell **F2**
On the **Formulas** tab, select **COUNTIF** under the
More Functions -> Statistical menu.

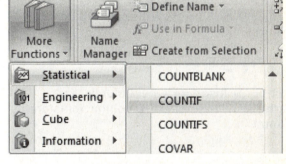

In the Function Arguments window,
Enter: **Range: $C$1:$C$100**
Enter: **Criteria: E2**
Click: **OK**

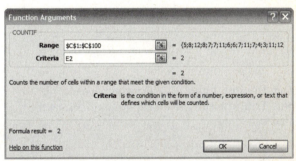

Remark: The use of absolute cell references will allow
us to copy this formula down column **F**. The
COUNTIF function tallies all of the values in column
C that equal the value found in cell **E2**.

Alternatively:
Enter: Formula **=COUNTIF($C$1:$C$100,E2)** in
cell **F2**.

Copy: Contents of cell **F2** to cells **F3:F12**

	A	B	C	D	E	F
1	1	4	5	Total	Frequency
2	1	3	4	2	0
3	5	2	7	3	4
4	4	4	8	4	10
5	5	3	8	5	7
6	2	5	7	6	9
7	5	5	10	7	19
8	5	6	11	8	19
9	3	5	8	9	12
10	5	5	10	10	11
11	6	5	11	11	8
12	1	5	6	12	1
13	6	3	9		

We now generate a bar graph illustrating these frequencies.

Select: Cells **F2:F12**

Select: On the **Insert** tab, select **Column** on the **Charts** menu. Then, select **2-D column (clustered column)**.

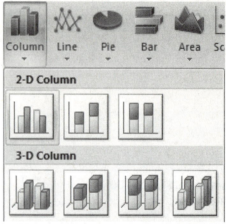

Select: **Design, Layout,** or **Format** tab under the **Chart Tools** menu. Use the submenus to format the output as desired.

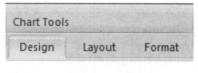

For example,
Select: **Design** tab
Select: **Select Data** on the Data submenu

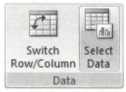

In **Select Data Source** window,
Select: **Horizontal (Category) Axis Labels: Edit**

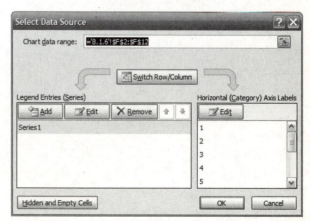

In the **Axis Labels** window,
Select: **Axis label range: E2:E12**
Click: **OK**

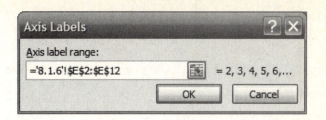

In **Select Data Source** window,
Click: **OK**

Remark: When entering the axis label range, use the mouse to highlight the appropriate cells. Note in this case that cells **E2:E12** are found on the Sheet labeled **8.1.6**.

For example,
Select: **Format** tab

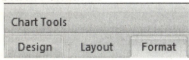

Select: **Series 1** on the **Current Selection** submenu

Select: **Format Selection** on the **Current Selection** submenu

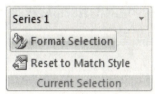

In the **Format Data Series** window,
Select: **Series Options**
Enter: **Gap Width: 25%**
Click: **Close**

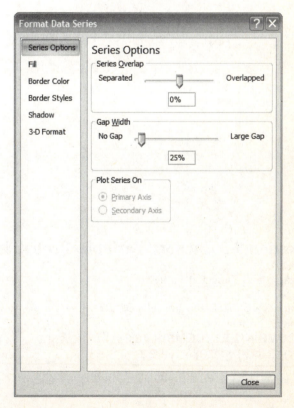

Sample output is shown to the right.

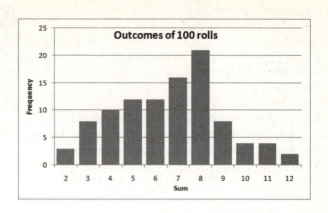

Section 8-2 Union, Intersection, and Complement of Events; Odds

Example 5: Birthday Problem

To construct a table like that in Table 1,
Enter: Value **5** in cell **A2**
Enter: Formula **=A2+1** in cell **A3**
Copy: Contents of cell **A3** to cells **A4:A47**
Enter: Formula **=1-PERMUT(365,A2)/(365^A2)**
in cell **B2**

	A	B	C	D
1	n	P(E)		
2	5	=1-PERMUT(365,A2)/(365^A2)		
3	6			
4	7			
5	8			
6	9			

Copy: Contents of cell **B2** to cells **B3:B47**

	A	B
1	n	P(E)
2	5	0.0271356
3	6	0.0404625
4	7	0.0562357
5	8	0.0743353
6	9	0.0946238
7	10	0.1169482
8	11	0.1411414
9	12	0.1670248
10	13	0.1944103
11	14	0.2231025
12	15	0.2529013
13	16	0.283604

Section 8-5 Random Variable, Probability Distribution, and Expected Value

Example 2: Expected Value

Enter: Values **0, 1,** and **2** in cells **B1:D1**
Enter: Formula
=COMBIN(2,B1)*COMBIN(18,3-B1)/COMBIN(20,3) in cell **B2**

	A	B	C	D	E	F	G
1	x_i	0	1	2		E(X)	
2	p_i	=COMBIN(2,B1)*COMBIN(18,3-B1)/COMBIN(20,3)					

Copy: Contents of cell **B2** to cells **C2:D2**

	A	B	C	D
1	x_i	0	1	2
2	p_i	0.715789	0.268421	0.015789

Enter: Formula **=B1*B2** in cell **B3**
Copy: Contents of cell **B3** to cells **C3:D3**

	A	B	C	D
1	x_i	0	1	2
2	p_i	0.715789	0.268421	0.015789
3		=B1*B2		

Enter: Formula **=SUM(B3:D5)** in cell **F5**

	A	B	C	D	E	F	G
1	x_i	0	1	2		E(X)	
2	p_i	0.715789	0.268421	0.015789			
3		0	0.268421	0.031579		=SUM(B3:D3)	

Chapter 9

Section 9-1 Properties of Markov Chains

Example 1: Insurance

Enter: Transition matrix P in cells **B1:C3**
Enter: Initial-state matrix S_0 in cells
B4:C4

	A	B	C
1	P	0.23	0.77
2		0.11	0.89
3			
4	S_0	0.05	0.95

Find the first state S_1:

Select: Cells **B6:C6**

Select: **MMULT** from the **Math & Trig**
menu under the **Formulas** tab.

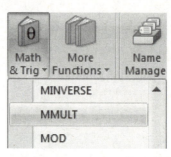

In the **Function Arguments** window,
Select: **Array1**: Cells **B4:C4**
Note: Use the mouse to select the cells in
which this matrix is located rather than
typing **B4:C4.**

Select: **Array2**: Cells **$B$1:$C$2**
Note: Using anchors ($) and making
absolute references will allow easy
copying for later calculations.

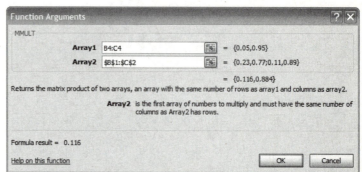

Do NOT click OK. Instead, use **CRTL-
SHIFT-ENTER** (press **ENTER** while
holding **CRTL** and **SHIFT** down) rather
than just ENTER to input an *array
formula.*
Alternatively:
Enter: The array formula
=MMULT(B4:C4, $B$1:$C$2) in
selected region **B6:C6** (not in just a single
cell)**.**

*Use CRTL-SHIFT-ENTER rather than
ENTER to input an array formula.

	A	B	C	D
1	P	0.23	0.77	
2		0.11	0.89	
3				
4	S_0	0.05	0.95	
5				
6	S_1	=MMULT(B4:C4,$B$1:$C$2)		

Find the second state S_2:

Copy: Contents of cells **B6:C6** to cells **B8:C8**

	A	B	C
1	P	0.23	0.77
2		0.11	0.89
3			
4	S_0	0.05	0.95
5			
6	S_1	0.116	0.884
7			
8	S_2	0.12392	0.87608

Example 2: Using P^k to Compute S_k

Enter: The transition matrix P in cells **B1:C2**

Select: Cells **B4:C5**
Enter: The array formula **=MMULT(B1:C2,B1:C2)** in selected region

Note: Using **MMULT** from the **Math & Trig** menu under the **Formulas** tab makes this relatively easy.

*Use CRTL-SHIFT-ENTER rather than ENTER to input an array formula.

	A	B	C	D
1	P	0.1	0.9	
2		0.6	0.4	
3				
4	P^2	=MMULT(B1:C2,B1:C2)		
5				

Copy: Contents of cells **B4:C5** to cells **B7:C8**

	A	B	C
1	P	0.1	0.9
2		0.6	0.4
3			
4	P^2	0.55	0.45
5		0.3	0.7
6			
7	P^4	0.4375	0.5625
8		0.375	0.625

Enter: The initial-state matrix S_0 in cells **E1:F1**
Select: Cells **E7:F7**
Enter: The array formula **=MMULT(E1:F1,B7:C8)** in selected region

*Use CRTL-SHIFT-ENTER rather than ENTER to input an array formula.

	A	B	C	D	E	F	G
1	P	0.1	0.9	S_0	0.2	0.8	
2		0.6	0.4				
3							
4	P^2	0.55	0.45				
5		0.3	0.7				
6							
7	P^4	0.4375	0.5625	S_4	=MMULT(E1:F1,B7:C8)		
8		0.375	0.625				

Example 4: University Enrollment

Enter: The transition matrix *P* in cells **B1:E4**

Select: Cells **B6:E9**
Enter: The array formula = **MMULT(B1:E4,B1:E4)**
in selected region

Note: Using **MMULT** from the **Math & Trig** menu
under the **Formulas** tab makes this relatively easy.

*Use CRTL-SHIFT-ENTER rather than ENTER to
input an array formula.

Copy: Contents of cells **B6:E9** to cells **B11:E14**

	A	B	C	D	E
1	P	0.6	0.1	0.3	0
2		0	1	0	0
3		0	0.1	0.5	0.4
4		0	0	0	1
5					
6	P^2	0.36	0.19	0.33	0.12
7		0	1	0	0
8		0	0.15	0.25	0.6
9		0	0	0	1

	A	B	C	D	E
1	P	0.6	0.1	0.3	0
2		0	1	0	0
3		0	0.1	0.5	0.4
4		0	0	0	1
5					
6	P^2	0.36	0.19	0.33	0.12
7		0	1	0	0
8		0	0.15	0.25	0.6
9		0	0	0	1
10					
11	P^4	0.1296	0.3079	0.2013	0.3612
12		0	1	0	0
13		0	0.1875	0.0625	0.75
14		0	0	0	1

Section 9-2 Regular Markov Chains

Example 1: Recognizing Regular Matrices

Enter: The transition matrix *P* in cells **B1:D3**

Select: Cells **B5:D7**
Enter: The array formula
=MMULT(B1:D3,$B$1:$D$3) in selected region

Note: Using **MMULT** from the **Math & Trig** menu
under the **Formulas** tab makes this relatively easy.

*Use CRTL-SHIFT-ENTER rather than ENTER to
input an array formula.

	A	B	C	D
1	P	0.5	0.5	0
2		0	0.5	0.5
3		1	0	0
4				
5	P^2	=MMULT(B1:D3,$B$1:$D$3)		
6				
7				

Copy: The contents of cells **B5:D7** to cells **B9:D11**

*Note the use of the absolute reference in computing P^3

	A	B	C	D
1	P	0.5	0.5	0
2		0	0.5	0.5
3		1	0	0
4				
5	P^2	0.25	0.5	0.25
6		0.5	0.25	0.25
7		0.5	0.5	0
8				
9	P^3	0.375	0.375	0.25
10		0.5	0.375	0.125
11		0.25	0.5	0.25

Example 2: Finding the Stationary Matrix

Enter: The transition matrix *P* in cells **B1:C2**

	A	B	C
1	P	0.7	0.3
2		0.2	0.8
3			
4	S	0.5	0.5
5			
6	SP	=MMULT(B4:C4,B1:C2)	

Enter: Any initial-state matrix *S* (e.g. **[0.5 0.5]**) in cells **B4:C4**
Select: Cells **B6:C6**
Enter: The array formula
=MMULT(B4:C4,B1:C2) in selected region (using **MMULT** from the **Math & Trig** menu under the **Formulas** tab makes this relatively easy)
*Use CRTL-SHIFT-ENTER rather than ENTER to input an array formula.

We seek *S* (having components which sum to 1) such that *SP* = *S*. To do this, we will minimize the difference between *SP* and *S*. That is, minimize the sum of the square of the differences between the components.
Enter: Formula **=SUM(B4:C4)** in cell **D4**

	A	B	C	D	E
1	P	0.7	0.3		
2		0.2	0.8		
3					
4	S	0.5	0.5	=SUM(B4:C4)	
5					
6	SP	0.45	0.55		

Enter: Formula **=B6-B4** in cell **B8**
Copy: Contents of cell **B8** to cell **C8**

	A	B	C	D
1	P	0.7	0.3	
2		0.2	0.8	
3				
4	S	0.5	0.5	1
5				
6	SP	0.45	0.55	
7				
8	Difference	-0.05	0.05	

Enter: Formula **=B8^2** in cell **B9**
Copy: Contents of cell **B9** to cell **C9**
Enter: Formula **=SUM(B9:C9)** in cell **D9**

	A	B	C	D	E
1	P	0.7	0.3		
2		0.2	0.8		
3					
4	S	0.5	0.5	1	
5					
6	SP	0.45	0.55		
7					
8	Difference	-0.05	0.05		
9	Diff Sq	0.0025	0.0025	=SUM(B9:C9)	

Select: **Data** tab

Select: **Solver** (if **Solver** is not present, see chapter 6 for instructions on adding this tool)

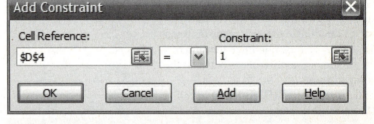

In **Solver Parameters** window,
Enter: **Select Target Cell: D9**
Select: **Equal To: Min**
Enter: **By Changing Cells: B4:C4**
Click: Subject to the Constraints: **Add**
Enter: Cell Reference: **D4**
Select: **=**
Enter: Constraint: **1**
Click: **OK**

In **Solver Parameters** window,
Click: **Solve**
Select: **Keep Solver Solution**
Click: **OK**

Since exactly one solution exists, an approximation to it appears in cells **B4:C4**.

	A	B	C	D
1	P	0.7	0.3	
2		0.2	0.8	
3				
4	S	0.4	0.6	1
5				
6	SP	0.4	0.6	
7				
8	Difference	0	0	
9	Diff Sq	0	0	0

Theorem 1 allows a different approach:
Enter: The transition matrix P in cells **B1:C2**

Select: Cells **B4:C5**
Enter: The array formula
=MMULT(B1:C2,$B$1:$C$2) in selected region
*Use CRTL-SHIFT-ENTER rather than ENTER to input an array formula.
Copy: Contents of cells **B4:C5** to cells **B7:C8**

Repeat as necessary. The resulting products will approach a matrix having rows consisting of the stationary matrix.

	A	B	C	D
1	P	0.7	0.3	
2		0.2	0.8	
3				
4	P^2	=MMULT(B1:C2,$B$1:$C$2)		
5				

	A	B	C
1	P	0.7	0.3
2		0.2	0.8
3			
4	P^2	0.55	0.45
5		0.3	0.7
6			
7	P^3	0.475	0.525
8		0.35	0.65
9			
10	P^4	0.4375	0.5625
11		0.375	0.625
12			
13	P^5	0.41875	0.58125
14		0.3875	0.6125
15			
16	P^6	0.409375	0.590625
17		0.39375	0.60625

Example 5: Approximating the Stationary Matrix

Enter: The transition matrix P in cells **B1:D3**

Select: Cells **B5:D7**
Enter: The array formula **=MMULT(B1:D3,B1:D3)** in selected region
*Use CRTL-SHIFT-ENTER rather than ENTER to input an array formula.

	A	B	C	D
1	P	0.5	0.2	0.3
2		0.7	0.1	0.2
3		0.4	0.1	0.5
4				
5	P^2	=MMULT(B1:D3,B1:D3)		
6				
7				

Copy: Contents of cells **B5:D7** to cells **B9:D11**

Repeat as necessary to approximate $\overline{P}$

◢	A	B	C	D
1	P	0.5	0.2	0.3
2		0.7	0.1	0.2
3		0.4	0.1	0.5
4				
5	P^2	0.51	0.15	0.34
6		0.5	0.17	0.33
7		0.47	0.14	0.39
8				
9	P^4	0.4949	0.1496	0.3555
10		0.4951	0.1501	0.3548
11		0.493	0.1489	0.3581
12				
13	P^8	0.494254	0.149426	0.35632
14		0.494256	0.149427	0.356317
15		0.494249	0.149424	0.356327

Chapter 10

Section 10-2 Mixed Strategy Games

Example 1: Solving a 2×2 Nonstrictly Determined Matrix Game

Enter: Game matrix M in cells **A2:B3**
Enter: Formula **=A2+B3-B2-A3** for D in cell **D2**
Enter: Formula **=(B3-A3)/D2** in cell **A6**
Enter: Formula **=(A2-B2)/D2** in cell **B6**
Enter: Formula **=(B3-B2)/D2** in cell **A9**
Enter: Formula **=(A2-A3)/D2** in cell **A10**
Enter: Formula **=(A2*B3-B2*A3)/D2** in cell **A13**

*Note: By Theorem 4, this setup will solve *all* 2×2 nonstrictly determined matrix games – simply change the values in cells **A2:B3**.

	A	B	C	D	E
1	M			D	
2	2	-3		=A2+B3-B2-A3	
3	-3	4			
4					
5	P*				
6	0.583333	0.416667			
7					
8	Q*				
9	0.583333				
10	0.416667				
11					
12	v				
13	-0.08333				

Section 10-3 Linear Programming and 2×2 Games: A Geometric Approach

Example 1: Solving 2×2 Matrix Games Using Geometric Methods

Enter: Game matrix M in cells **A2:B3**

Convert M to a positive matrix:
Enter: Formula
=A2+MAX($A$2:$B$3) in cell **D2**
Copy: Contents of cell **D2** to cells **D2:E3**
Enter: Any values (e.g. **0**) in cells **A6** and **B6** (representing the values of x_1 and x_2)
Enter: Formula **=A6+B6** in cell **D6** (representing the value of y which we try to minimize)
We now enter the constraints:
Enter: Formula **=D2*A6+D3*B6** in cell **A9**
Enter: Value **1** in cell **B9**
Enter: Formula **=E2*A6+E3*B6** in cell **A10**
Enter: Value **1** in cell **B10**

	A	B	C	D	E	F
1	M			M 1		
2	-2	4		=A2+MAX($A$2:$B$3)		
3	1	-3				
4						

	A	B	C	D	E
1	M			M_1	
2	-2	4		2	8
3	1	-3		5	1
4					
5	x_1	x_2		y	
6	0	0		=A6+B6	
7					
8	LHS	RHS			
9	0	1			
10	0	1			

Select: **Data** tab

Formulas	Data	Review

Select: **Solver** (if **Solver** is not present, see chapter 6 for instructions on adding this tool)

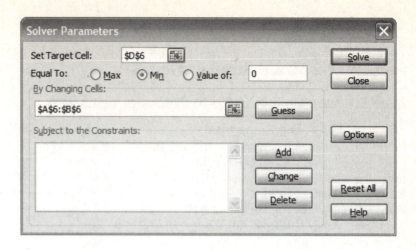

In **Solver Parameters** window,
Enter: **Select Target Cell: D6**
Select: **Equal To: Min**
Enter: **By Changing Cells: A6:B6**
Click: Subject to the Constraints: **Add**

Enter: Cell Reference: **A9**
Select: **>=**
Enter: Constraint: **B9**
Click: **Add**

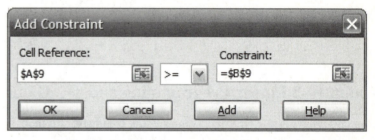

Enter: Cell Reference: **A10**
Select: **>=**
Enter: Constraint: **B10**
Click: **Add**

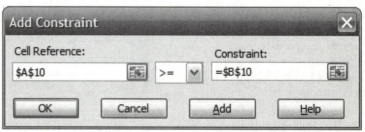

Enter: Cell Reference: **A6**
Select: **>=**
Enter: Constraint: **0**
Click: **Add**
Enter: Cell Reference: **B6**
Select: **>=**
Enter: Constraint: **0**
Click: **OK**
Click: **Solve**

In the **Solver Results** window,
Select: **Keep Solver Solution**
Click: **OK**

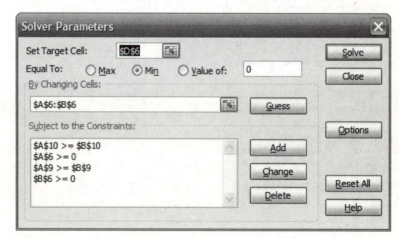

The solution (for the values of x_1 and x_2) will appear in cells **A6** and **B6**.

	A	B	C	D	E
1	M			M_1	
2	-2	4		2	8
3	1	-3		5	1
4					
5	x_1	x_2		y	
6	0.105263	0.157895		0.263158	
7					
8	LHS	RHS			
9	1	1			
10	1	1			

Repeat the above steps on the corresponding maximization problem.

Enter: Any values (e.g. **0**) in cells **A13** and **B13** (representing the values of z_1 and z_2)

Enter: Formula **=A13+B13** in cell **D13** (representing the value of y which we try to maximize)

Then, enter the constraints:
Enter: Formula **=D2*A13+E2*B13** in cell **A16**
Enter: Value **1** in cell **B16**
Enter: Formula **=D3*A13+E3*B13** in cell **A17**
Enter: Value **1** in cell **B17**

	A	B	C	D	E
1	M			M_1	
2	-2	4		2	8
3	1	-3		5	1
4					
5	x_1	x_2		y	
6	0.105263	0.157895		0.263158	
7					
8	LHS	RHS			
9	1	1			
10	1	1			
11					
12	z_1	z_2		y	
13	0	0		=A13+B13	
14					
15	LHS	RHS			
16	0	1			
17	0	1			

Select: **Data** tab

Select: **Solver**
In **Solver Parameters** window,
Enter: **Select Target Cell: D13**
Select: **Equal To: Max**
Enter: **By Changing Cells: A13:B13**
Click: Subject to the Constraints: **Add**

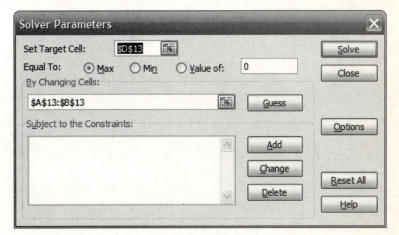

Enter the constraints shown at the right following the same process as before.

Click: **Solve**

In the **Solver Results** window, Select: **Keep Solver Solution** Click: **OK**

Enter: Formula **=1/D6** in cell **B19** (representing the value of v_1)

Determine the values of P^* and Q^*:
Enter: Formula **=B19*A6** in cell **A22**
Enter: Formula **=B19*B6** in cell **B22**
Enter: Formula **=B19*A13** in cell **D21**
Enter: Formula **=B19*B13** in cell **D22**

The value v of the original game is, in this case, 4 less than the value of v_1.

Enter: Formula **=B19-MAX($A$2:$B$3)** in cell **B24**

	A	B	C	D	E
1	M			M_1	
2	-2	4		2	8
3	1	-3		5	1
4					
5	x_1	x_2		y	
6	0.105263	0.157895		0.263158	
7					
8	LHS	RHS			
9	1	1			
10	1	1			
11					
12	z_1	z_2		y	
13	0.184211	0.078947		0.263158	
14					
15	LHS	RHS			
16	1	1			
17	1	1			
18					
19	v_1	3.8			
20				Q*	
21	P*			0.7	
22	0.4	0.6		0.3	
23					
24	v	=B19-MAX($A$2:$B$3)			

Chapter 11

Section 11-1 Graphing Data

Figure 1: Creating a vertical bar graph

Enter: Data from Table 1

	A	B
1	Year	Debt (billions $)
2	1968	348
3	1978	772
4	1988	2602
5	1998	5526
6	2008	10699

Select: Cells **B2:B6**

Select: On the **Insert** tab, select **Column** on the **Charts** menu. Then, select **2-D (Clustered) Column**.

Select: **Design** tab under the **Chart Tools** menu.

Select: **Select Data** on the **Data** submenu.

In the **Select Data Source** window,

Click: **Edit** button under **Horizontal (Category) Axis Labels**

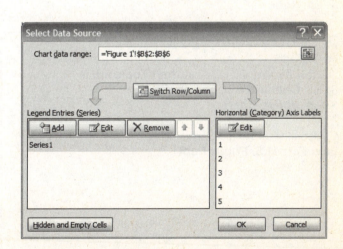

In the **Axis Labels** window,

Select: **Axis label range** to be **A2:A6**.
Click: **OK**

Good Habit: Use the mouse to select these cells rather than typing **A2:A6** in this window from the keyboard.

Output will look as at the right. We are now ready to format it to our specification.

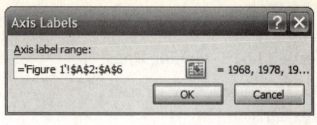

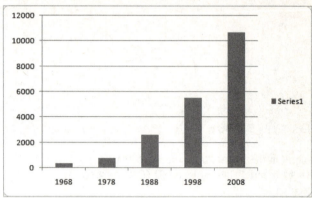

For example,
Select: **Layout** tab under the **Chart Tools** menu.

Select: **Legend** on the **Labels** menu.
Select: **None**

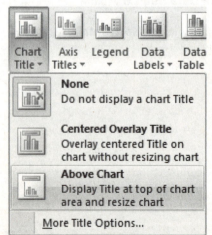

For example,
Select: **Layout** tab under the **Chart Tools** menu.
Select: **Gridlines** on the **Axes** menu.
Select: **Primary Horizontal Gridlines**
Select: **None**

For example,
Select: **Layout** tab under the **Chart Tools** menu.
Select: **Chart Title** on the **Labels** menu.
Select: **Above Chart**

Click on the title in the graph.
Enter Text: **U.S. Public Debt**

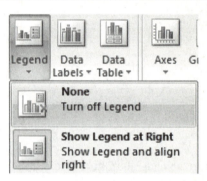

For example,
Select: **Layout** tab under the **Chart Tools** menu.
Select: **Axis Titles** on the **Labels** menu.
Select: **Primary Horizontal Axis Title**
Select: **Title Below Axis**

Click on the title below the axis.
Enter Text: **Year**

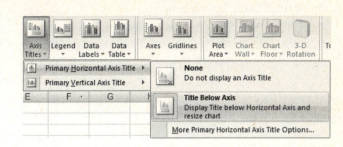

Select: **Layout** tab under the **Chart Tools** menu.
Select: **Axis Titles** on the **Labels** menu.
Select: **Primary Vertical Axis Title**
Select: **Rotated Title**

Click on the title appearing on the vertical axis.
Enter Text: **Billion Dollars**

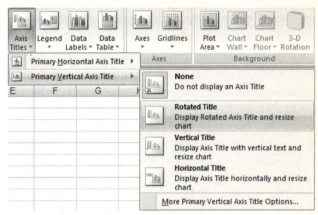

For example,
Select: **Format** tab under the **Chart Tools** menu.
Select: **Series 1** in the **Current Selection**
window.
Select: **Format Selection** from the **Current
Selection window**.

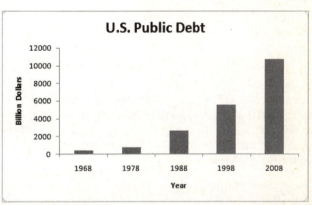

Alternatively,
On the graph, use the mouse to select any of the
bars. Right click to select series.
Select: **Format Data Series**

In the **Format Data Series** window,
Select: **Series Options**
Select: **Gap Width: 50%**
Click: **Close**

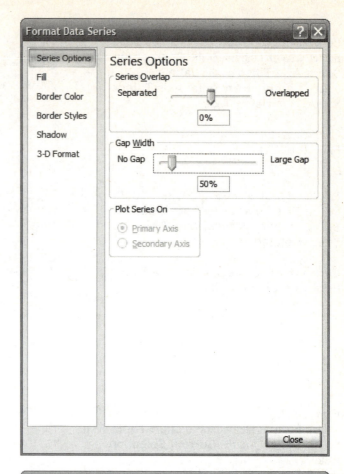

For example,
Select: **Format** tab under the **Chart Tools** menu.
Select: **Vertical (Value) Axis** in the **Current Selection** window.
Select: **Format Selection** from the **Current Selection window**.

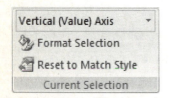

In the **Format Axis** window,
Select: **Number**
Select: **Category: Currency**
Enter: **Decimal places: 0**

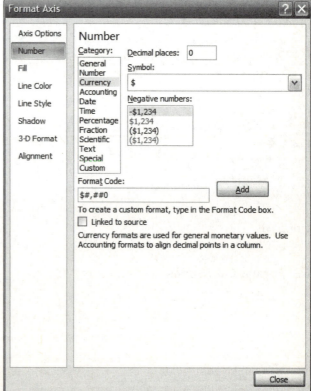

Sample output is shown at the right.

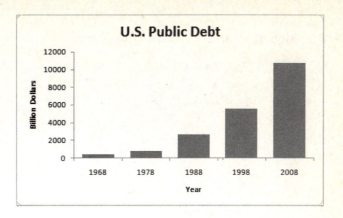

Figure 2: Creating a horizontal bar graph

Enter: Data from Table 2

	A	B
1	**Airport**	**Arrivals and Departures**
2	Atlanta	89
3	Chicago (O'Hare)	76
4	Los Angeles	62
5	Dallas/Ft. Worth	60
6	Denver	50

Select: Cells **A2:B6**

Select: On the **Insert** tab, select **Bar** on the **Charts** menu. Then, select **2-D (Clustered) Bar**.

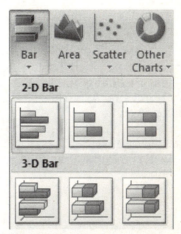

Output will look as at the right. We are now ready to format it to our specification.

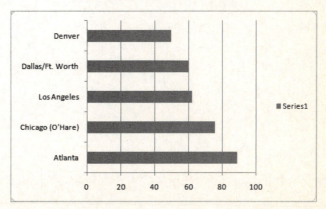

For example,
Select: **Layout** tab under the **Chart Tools** menu.

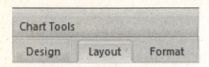

Select: **Legend** on the **Labels** menu.
Select: **None**

For example,
Select: **Layout** tab under the **Chart Tools** menu.
Select: **Chart Title** on the **Labels** menu.
Select: **Above Chart**

Click on the title in the graph.
Enter Text: **Traffic at Busiest U.S. Airports, 2005**

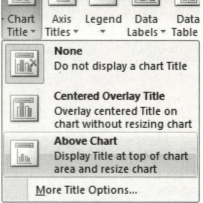

For example,
Select: **Layout** tab under the **Chart Tools** menu.
Select: **Axis Titles** on the **Labels** menu.
Select: **Primary Horizontal Axis Title**
Select: **Title Below Axis**

Click on the title below the axis.
Enter Text: **Arrivals and Departures (million passengers)**

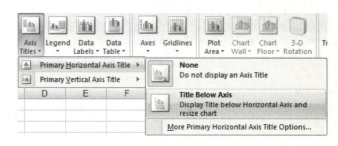

For example,
Select: **Format** tab under the **Chart Tools** menu.
Select: **Series 1** in the **Current Selection** window.
Select: **Format Selection** from the **Current Selection window**.

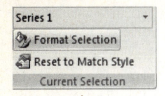

In the **Format Data Series** window,
Select: **Series Options**
Select: **Gap Width: 100%**

In the **Format Data Series** window,
Select: **Fill**
Select: **Gradient fill**
Select: **Type: Linear**
Select: **Angle: 180 degrees**
Select: **Preset colors: Your choice!**
Select: **Close**

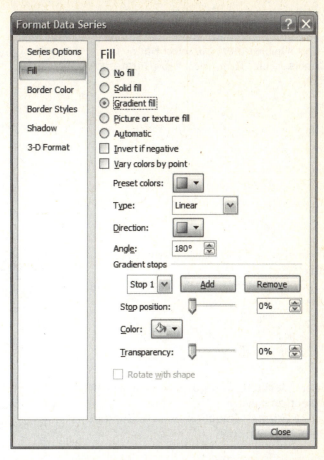

Sample output is shown at the right.

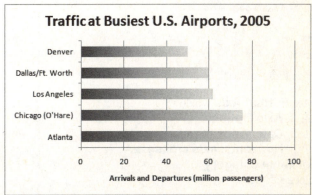

Figure 3: Creating a double bar graph

Enter: Data for the graph

	A	B	C
1	Education	Male Income	Female Income
2	Bachelor's degree	$60,910	$45,410
3	Associate degree	$47,070	$36,160
4	Some college	$43,830	$31,950
5	High school diploma	$37,030	$26,740
6	Some high school	$27,650	$20,130

Select: Cells **A2:C6**

Select: On the **Insert** tab, select **Bar** on the
Charts menu. Then, select **2-D (Clustered)
Bar**.

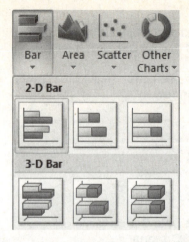

The output can be formatted to have the desired
effect. For example,
Select: **Layout** tab under the **Chart Tools** menu.

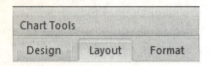

Select: **Vertical (Category) Axis** in the **Current
Selection** window.
Select: **Format Selection** from the **Current
Selection window**.

In the **Format Axis** window,
Select: **Axis Options**
Select: **Categories in reverse order**
Select: **Horizontal axis crosses: At
maximum category**
Select: **Close**

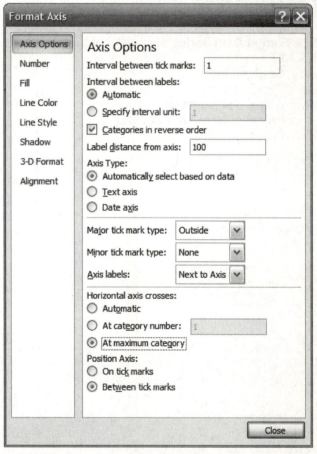

For example,
Select: **Layout** tab under the **Chart Tools** menu.
Select: **Gridlines** on the **Axes** menu.
Select: **Primary Vertical Gridlines**
Select: **More Primary Vertical Gridlines Options...**

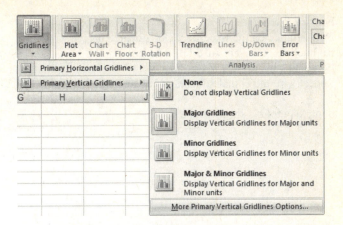

In the **Format Major Gridlines** window,
Select: **Line Style**
Select: **Width: 1.5 pt**
Select: **Dash type: Round Dot**
Select: **Close**

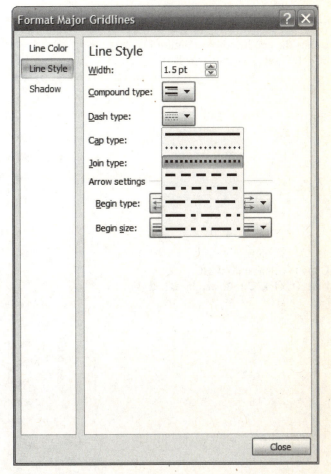

For example,
Double-click on the Male Bachelor's degree bar.
Right click and select **Add Data Label**
Edit data label to say "Male"
Repeat with the Female Bachelor's degree bar.

Sample output is shown at the right.

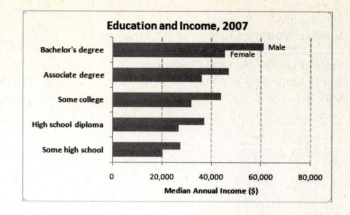

Figure 5: Creating a broken-line graph

Create the graph shown in Figure 1.
Select: **Chart** (click on the chart).

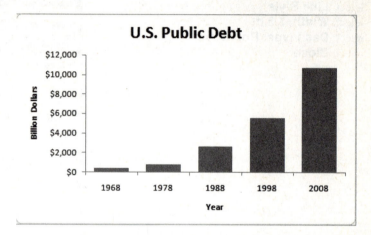

Copy the chart onto itself.
Enter: **CTRL-C** followed by **CTRL-V**

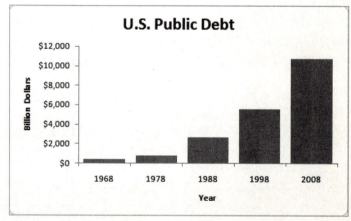

Right Click: Second set of bars
Select: **Change Series Chart Type...**

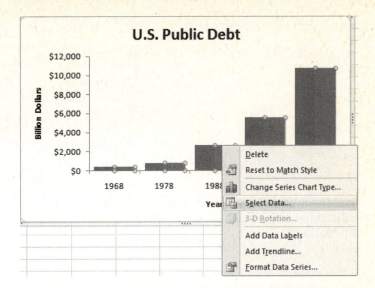

In the **Change Chart Type** window,
Select: **Line**
Click: **OK**

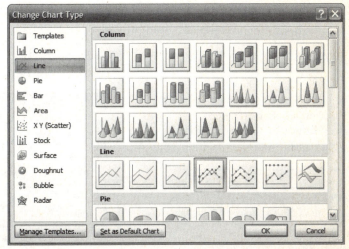

The result is called a broken-line graph.

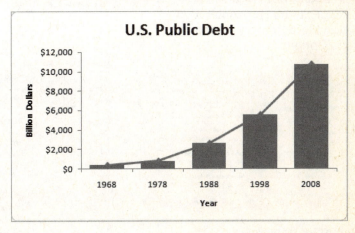

Figure 7: More Broken-Line Graphs

Enter: Data for the line graphs.

	A	B	C	D	E	F
1		**2010**	**2015**	**2020**	**2025**	**2030**
2	**Petroleum**	38.2	38.7	38.7	39.4	40.3
3	**Natural Gas**	22.5	21.6	22.1	23.9	24.2
4	**Coal**	22.3	23.7	24.4	24.5	25.4
5	**Nuclear**	8.5	8.7	9.1	9.2	9.3
6	**Renewable**	6.7	8.4	9	9.5	9.7

Select: Cells **A1:F6**

Select: On the **Insert** tab, select **Line** on the **Charts** menu. Then, select **Line With Markers**.

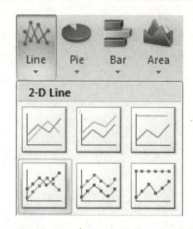

The output can be formatted to have the desired effect. For example,
Select: **Layout** tab under the **Chart Tools** menu.

Select: **Axes** on the **Axes** menu.
Select: **Primary Horizontal Axis**
Select: **More Primary Horizontal Axis Options…**

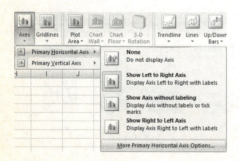

In **Format Axis** window,
Select: **Axis Options**
Select: **Position Axis: On tick marks**
Select: **Close**

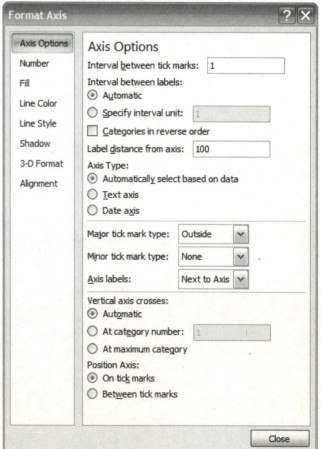

For example,
Select: **Layout** tab under the **Chart Tools** menu.

Select: The last data point for a particular line in the graph (by double-clicking)
Select: **Data Labels** from the **Labels** menu
Select: **More Data Labels**

In the **Format Data Labels** window,
Select: **Label Contains: Series Name**
Select: **Close**

Sample output is shown at the right.

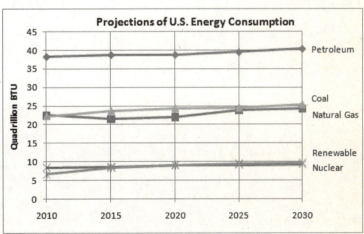

Figure 8: Creating a pie graph

Enter: Data for the pie graph

	A	B
1	**Army**	519471
2	**Navy**	338671
3	**Air Force**	337312
4	**Marine Corps**	184574

Select: Cells **A1:B4**

Select: On the **Insert** tab, select **Pie** on the **Charts** menu. Then, select **2-D Pie**.

Format the data as desired.

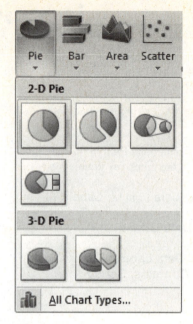

For example,
Select: **Layout** tab under the **Chart Tools** menu.

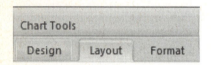

Select: **Data Labels** on **Labels** menu
Select: **Center**

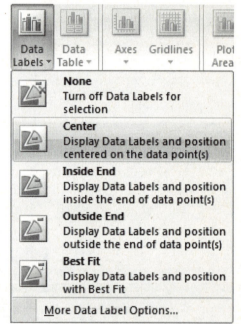

Select: **Layout** tab under the **Chart Tools** menu.

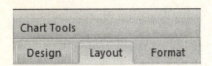

Select: **Series 1 Data Labels** in the **Current Selection** window.
Select: **Format Selection** from the **Current Selection window**.

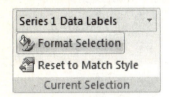

In the **Format Data Labels** window,
Select: **Label Options**
Select: **Label Contains: Category Name**
Select: **Label Contains: Percentage**
Select: **Label Position: Center**

Select: **Number**
Select: **Category: Percentage**
Select: **Decimal Places: 0**
Select: **Close**

Sample output is shown at the right.

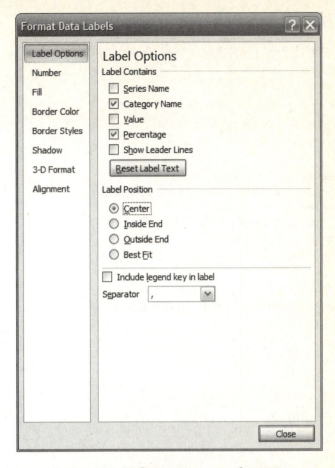

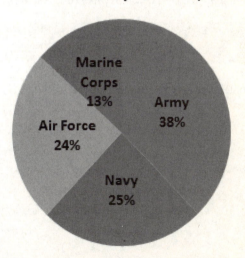

Active U.S. Military Personnel, 2007

Figure 11: Creating a frequency table (and histogram)

Enter: Data (from Table 3) in cells **A1:A100**

	A
1	762
2	433
3	712
4	566
5	618
6	340

Set up the class intervals.
Enter: Value **299.5** in cell **C2** and value **349.5** in cell **D2**
Enter: Formula **=D2** in cell **C3** and formula **=D2+50** in cell **D3**

Copy: Contents of cells **C3:D3** to cells **C4:D11**

Good Habit: Click and hold the left mouse button on the lower left corner of cell **D3** and drag this down to cell **D11**.

	A	B	C	D
1	762		Class Interval	
2	433		299.5	349.5
3	712		349.5	=D2+50

	A	B	C	D
1	762		Class Interval	
2	433		299.5	349.5
3	712		349.5	399.5
4	566		399.5	449.5
5	618		449.5	499.5
6	340		499.5	549.5
7	548		549.5	599.5
8	442		599.5	649.5
9	663		649.5	699.5
10	544		699.5	749.5
11	451		749.5	799.5

Enter: Formula **=COUNTIF($A$1:$A$100,"<="&D2)-COUNTIF($A$1,$A$100,"<"&C2)** in cell **F2**

	A	B	C	D	E	F	G	H	I	J	K
1	762		Class Interval			Frequency					
2	433		299.5	349.5		=COUNTIF($A$1:$A$100,"<="&D2)-COUNTIF($A$1:$A$100,"<"&C2)					

Copy: Contents of cell **F2** to cells **F3:F11**

	A	B	C	D	E	F
1	762		Class Interval			Frequency
2	433		299.5	349.5		1
3	712		349.5	399.5		2
4	566		399.5	449.5		5
5	618		449.5	499.5		10
6	340		499.5	549.5		21
7	548		549.5	599.5		20
8	442		599.5	649.5		19
9	663		649.5	699.5		11
10	544		699.5	749.5		7
11	451		749.5	799.5		4

Note: Frequency tables and histograms can be constructed using the Histogram option on the Data Analysis tool However, Excel 2007 does a sloppy job of properly labeling the *x*-axis and a few tricks are necessary to deal with this problem. A simple search of the web will reveal the (cumbersome!) tricks that can be used to create a histogram such as that shown in Figure 11.

Output shown at the right is almost as good as one can do using Excel.

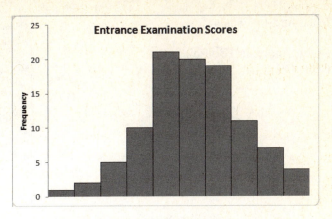

OPTIONAL Histogram construction process:

The **Analysis ToolPak** Add-in for Excel can help plot histograms. *If this Add-in has not been installed before*, you will need to install it.

Select: **Office Button** (shown at right)
Select: **Excel Options** button at bottom

In the **Excel Options** window,

Select: **Add-Ins** on the left menu
Select: **Analysis Toolpak**
Click: **Go…**

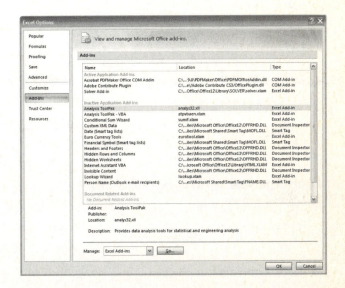

In the **Add-Ins** window,

Select: **Analysis ToolPak**
Click: **OK**

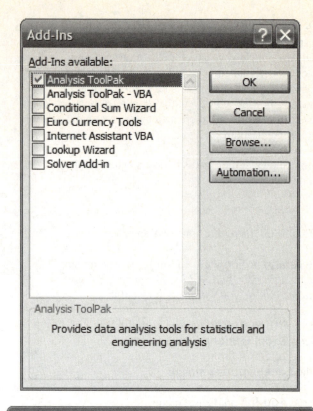

In the **Microsoft Office Excel** window,

Select: **Yes**

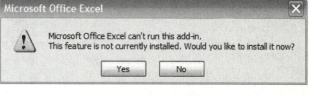

The **Data Analysis** tool will appear under the **Data** tab on the menu.

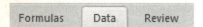

Click: **Data Analysis** button

In the Data Analysis window,
Select: **Histogram**
Click: **OK**

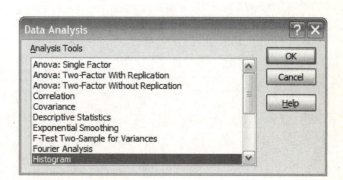

Enter: **Input Range: A1:A100** (where the data is stored)
Enter: Desired **Bin Range**
Select: **OK**

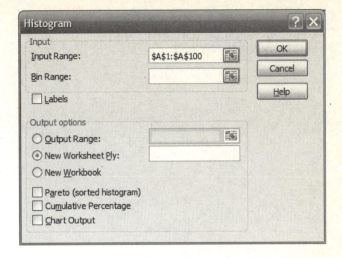

Figure 13: **Creating a frequency polygon**

Construct a frequency table in the same manner as described earlier. Add a class interval to the left and right (both having frequency of 0).

	A	B	C	D	E	F
1	762		Class Interval			Frequency
2	433		249.5	299.5		0
3	712		299.5	349.5		1
4	566		349.5	399.5		2
5	618		399.5	449.5		5
6	340		449.5	499.5		10
7	548		499.5	549.5		21
8	442		549.5	599.5		20
9	663		599.5	649.5		19
10	544		649.5	699.5		11
11	451		699.5	749.5		7
12	508		749.5	799.5		4
13	415		799.5	849.5		0

Enter: Formula **=(C2+D2)/2** in cell **E2**

	A	B	C	D	E	F
1	762		Class Interval		Midpoint	Frequency
2	433		249.5	299.5	=(C2+D2)/2	

Copy: Contents of cell **E2** to cells **E3:E13**

Select: Cells **E2:F13**

	A	B	C	D	E	F
1	762		Class Interval		Midpoint	Frequency
2	433		249.5	299.5	274.5	0
3	712		299.5	349.5	324.5	1
4	566		349.5	399.5	374.5	2
5	618		399.5	449.5	424.5	5
6	340		449.5	499.5	474.5	10
7	548		499.5	549.5	524.5	21
8	442		549.5	599.5	574.5	20
9	663		599.5	649.5	624.5	19
10	544		649.5	699.5	674.5	11
11	451		699.5	749.5	724.5	7
12	508		749.5	799.5	774.5	4
13	415		799.5	849.5	824.5	0

Select: On the **Insert** tab, select **Scatter** on the
Charts menu. Then, select **Scatter with
Straight Lines and Markers**.

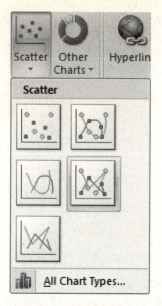

Format the output as desired.
For example,
Select: **Layout** tab under the **Chart Tools** menu.

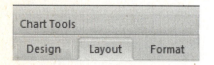

Select: **Horizontal (Value) Axis** in the **Current
Selection** window.
Select: **Format Selection** from the **Current
Selection window**.

In the **Format Axis** window,
Select: **Axis Options: Minimum: Fixed: 274.5**
Select: **Axis Options: Maximum: Fixed: 824.5**
Select: **Axis Options: Major unit: Fixed: 50**
Select: **Close**

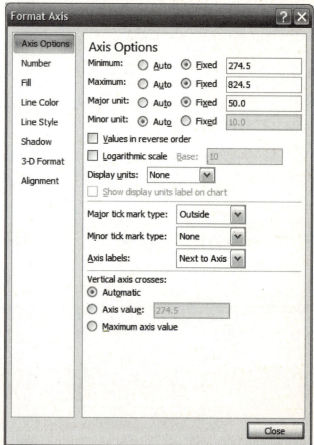

Sample output is shown at the right.

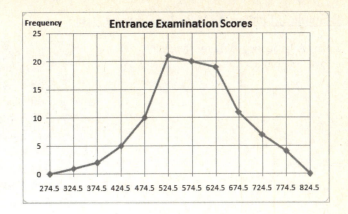

Figure 14: Creating a cumulative frequency polygon (table)

Start with the frequency table constructed earlier.

Enter: Formula **=SUM($F$2:F2)** to cell **G2**

	A	B	C	D	E	F	G
1	762		Class Interval		Midpoint	Frequency	Cumulative Frequency
2	433		249.5	299.5	274.5	0	=SUM($F$2:F2)

Copy: Contents of cell **G2** to cells **G3:G13**
To construct the cumulative frequency polygon, proceed as in the previous example using values **G2:G13** instead of values **F2:F13.**

	A	B	C	D	E	F	G
1	762		Class Interval		Midpoint	Frequency	Cumulative Frequency
2	433		249.5	299.5	274.5	0	0
3	712		299.5	349.5	324.5	1	1
4	566		349.5	399.5	374.5	2	3
5	618		399.5	449.5	424.5	5	8
6	340		449.5	499.5	474.5	10	18
7	548		499.5	549.5	524.5	21	39
8	442		549.5	599.5	574.5	20	59
9	663		599.5	649.5	624.5	19	78
10	544		649.5	699.5	674.5	11	89
11	451		699.5	749.5	724.5	7	96
12	508		749.5	799.5	774.5	4	100
13	415		799.5	849.5	824.5	0	100

Section 11-2 Measures of Central Tendency

Example 1: Finding the Mean

Enter: Data in cells **A1:A8**
Enter: Formula **=AVERAGE(A1:A8)** in cell **C2**

	A	B	C	D
1	3		Mean	
2	5		=AVERAGE(A1:A8)	
3	1			
4	8			
5	6			
6	5			
7	4			
8	6			

Example 2: Finding the Mean for Grouped Data

Enter: A frequency table for the data
as discussed in Section 11-1

	A	B	C
1	Class Interval	Midpoint x_i	Frequency f_i
2	299.5-349.5	324.5	1
3	349.5-399.5	374.5	2
4	399.5-449.5	424.5	5
5	449.5-499.5	474.5	10
6	499.5-549.5	524.5	21
7	549.5-599.5	574.5	20
8	599.5-649.5	624.5	19
9	649.5-699.5	674.5	11
10	699.5-749.5	724.5	7
11	749.5-799.5	774.5	4

Enter: Formula **=B2*C2** in cell **D2**
Copy: Contents of cell **D2** to cells
D3:D11

	A	B	C	D
1	Class Interval	Midpoint x_i	Frequency f_i	Product x_i*f_i
2	299.5-349.5	324.5	1	324.5
3	349.5-399.5	374.5	2	749
4	399.5-449.5	424.5	5	2122.5
5	449.5-499.5	474.5	10	4745
6	499.5-549.5	524.5	21	11014.5
7	549.5-599.5	574.5	20	11490
8	599.5-649.5	624.5	19	11865.5
9	649.5-699.5	674.5	11	7419.5
10	699.5-749.5	724.5	7	5071.5
11	749.5-799.5	774.5	4	3098

Enter: Formula
=SUM(D2:D11)/SUM(C2:C11) in
cell **A14**

	A	B	C	D
1	Class Interval	Midpoint x_i	Frequency f_i	Product x_i*f_i
2	299.5-349.5	324.5	1	324.5
3	349.5-399.5	374.5	2	749
4	399.5-449.5	424.5	5	2122.5
5	449.5-499.5	474.5	10	4745
6	499.5-549.5	524.5	21	11014.5
7	549.5-599.5	574.5	20	11490
8	599.5-649.5	624.5	19	11865.5
9	649.5-699.5	674.5	11	7419.5
10	699.5-749.5	724.5	7	5071.5
11	749.5-799.5	774.5	4	3098
12				
13	Mean			
14	=SUM(D2:D11)/SUM(C2:C11)			

Example 3: Finding the Median

Enter: Data in cells **A1:A7**
Enter: Formula **=MEDIAN(A1:A7)** in cell **A10**

Note: The data need not be ordered in order to use
the built-in **MEDIAN** function.

	A	B
1	$34,000	
2	$36,000	
3	$36,000	
4	$40,000	
5	$48,000	
6	$56,000	
7	$156,000	
8		
9	Median	
10	=MEDIAN(A1:A7)	

Example 5: Finding Mode, Median, and Mean

Enter: Data in cells **A2:A10**
Enter: Formula **=MODE(A2:A10)** in cell **B2**
Enter: Formula **=MEDIAN(A2:A10)** in cell **C2**
Enter: Formula **=AVERAGE(A2:A10)** in cell **D2**

	A	B	C	D
1	Data	Mode	Median	Mean
2	4	=MODE(A2:A10)		
3	5			
4	5			
5	5			
6	6			
7	6			
8	7			
9	8			
10	12			

Note: If a data set has more than one mode (such as
the one at the right having mode(s) of 3 and 7),
MODE only returns one of these.

Note: If the data set contains no duplicate data
points, MODE returns the #N/A error value.

	A	B	C	D
1	Data	Mode	Median	Mean
2	1	3	5	6.090909
3	2			
4	3			
5	3			
6	3			
7	5			
8	6			
9	7			
10	7			
11	7			
12	23			

Section 11-3 Measures of Dispersion

<u>**Example 1: Finding the Standard Deviation**</u>

Enter: Data in cells **A2:A6**
Enter: Formula **=A2-AVERAGE($A$2-$A$6)** in cell **B2**.
Copy: Contents in cell **B2** to cells **B3:B6**. This subtracts the deviation of each sample measurement from the mean.

	A	B	C	D
1	Data	Dev. Mean	Squared Dev	Sample Variance
2	1	=A2-AVERAGE($A$2:$A$6)		
3	3			
4	5			Sample Std Dev
5	4			
6	3			

Enter: Formula **=B2^2** in cell **C2**.
Copy: Contents in cell **C2** to cells **C3:C6**. This squares the deviations of each sample measurement from the mean.

	A	B	C	D
1	Data	Dev. Mean	Squared Dev	Sample Variance
2	1	-2.2	=B2^2	
3	3	-0.2		
4	5	1.8		Sample Std Dev
5	4	0.8		
6	3	-0.2		

Enter: Formula **=SUM(C2:C6)/(5-1)** in cell **D2**. This is the sample variance.

	A	B	C	D
1	Data	Dev. Mean	Squared Dev	Sample Variance
2	1	-2.2	4.84	=SUM(C2:C6)/(5-1)
3	3	-0.2	0.04	
4	5	1.8	3.24	Sample Std Dev
5	4	0.8	0.64	
6	3	-0.2	0.04	

Enter: Formula **=SQRT(D2)** in cell **D5**. This is the sample standard deviation.

Remark: Remember that the built-in square root function can also be found under the **Formulas** tab in the **Math & Trig Function Library**.

	A	B	C	D
1	Data	Dev. Mean	Squared Dev	Sample Variance
2	1	-2.2	4.84	2.2
3	3	-0.2	0.04	
4	5	1.8	3.24	Sample Std Dev
5	4	0.8	0.64	=SQRT(D2)
6	3	-0.2	0.04	

Alternatively, Excel has a single built-in function **STDEV** for computing the sample standard deviation.

Enter: Formula **=VAR(A2:A6)** in cell **E2**. Enter: Formula **=STDEV(A2:A6)** in cell **E5**.

	A	B	C	D	E
1	Data	Dev. Mean	Squared Dev	Sample Variance	Sample Variance
2	1	-2.2	4.84	2.2	2.2
3	3	-0.2	0.04		
4	5	1.8	3.24	Sample Std Dev	Sample Std Dev
5	4	0.8	0.64	1.483239697	=STDEV(A2:A6)
6	3	-0.2	0.04		

Remark: Computing the population variance and standard deviation can be done using the **VARP** and **STDEVP** functions.

Example 2: Finding the Standard Deviation for Grouped Data

Enter: Data in cells **A2:A6** and their corresponding frequencies in cells **B2:B6**
Enter: Formula **=A2-AVERAGE($A$2-$A$6)** in cell **C2.**
Copy: Contents in cell **C2** to cells **C3:C6.**

	A	B	C	D	E
1	Data	Frequency	Dev Mean	Squared Dev	Sample Variance
2	8	1	=A2-AVERAGE($A$2:$A$6)		
3	9	2			
4	10	4			Sample Std Dev
5	11	2			
6	12	1			

Enter: Formula **=B2*C2^2** in cell **D2.**
Copy: Contents in cell **D2** to cells **D3:D6.**
The sum of the contents of column **D** is the sum of the squared deviations of the data set.

	A	B	C	D	E
1	Data	Frequency	Dev Mean	Squared Dev	Sample Variance
2	8	1	-2	=B2*C2^2	
3	9	2	-1		
4	10	4	0		Sample Std Dev
5	11	2	1		
6	12	1	2		

Enter: Formula **=SUM(D2:D6)/(SUM(B2:B6)-1)** in cell **E2.**
This is the sample variance.

	A	B	C	D	E	F	G
1	Data	Frequency	Dev Mean	Squared Dev	Sample Variance		
2	8	1	-2	4	=SUM(D2:D6)/(SUM(B2:B6)-1)		
3	9	2	-1	2			
4	10	4	0	0	Sample Std Dev		
5	11	2	1	2			
6	12	1	2	4			

Enter: Formula **=SQRT(E2)** in cell **E5.** This is the sample standard deviation.

	A	B	C	D	E
1	Data	Frequency	Dev Mean	Squared Dev	Sample Variance
2	8	1	-2	4	1.333333333
3	9	2	-1	2	
4	10	4	0	0	Sample Std Dev
5	11	2	1	2	=SQRT(E2)
6	12	1	2	4	

Section 11-4 Bernoulli Trials and Binomial Distributions

Example 3: Probability of *x* Successes in *n* Bernoulli Trials

Enter: Values for *n* (number of repeated Bernoulli trials), *p* (probability of success) in cells **A2** and **B2.**
Enter: Value **5** in cell **A2.**
Enter: Value **=1/6** in cell **B2.**
Enter: Formula **=1-B2** in cell **C2.**

	A	B	C
1	n	p	q
2	5	0.166666667	=1-B2

Enter: Value **0** in cell **A5**
Enter: Formula **=A5+1** in cell **A6**
Copy: Contents of cell **A6** to cells **A7:A10**
Enter: Formula **=COMBIN($A$2,A5)*$B$2^A5*$C$2^($A$2-A5)** in cell **B5.**

	A	B	C	D	E
1	n	p	q		
2	5	0.166666667	0.833333		
3					
4	x	Pr(x successes)			
5	0	=COMBIN($A$2,A5)*$B$2^A5*$C$2^($A$2-A5)			
6	1				
7	2				
8	3				
9	4				
10	5				

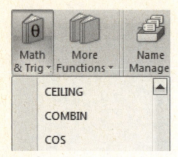

Remark: Remember that the built-in **COMBIN** function can also be found under the **Formulas** tab in the **Math & Trig Function Library**.

Copy: Contents of cell **B5** to cells **B6:B10**

The probability of exactly two successes can be found in cell **B7**. To compute the probability of at least two successes, use the formula **=SUM(B7:B10)** or the formula **=1-SUM(B5:B6)**.

	A	B	C
1	n	p	q
2	5	0.166666667	0.833333
3			
4	x	Pr(x successes)	
5	0	0.401877572	
6	1	0.401877572	
7	2	0.160751029	
8	3	0.032150206	
9	4	0.003215021	
10	5	0.000128601	

Example 7: Patient Recovery

Enter: Values for *n* (number of repeated Bernoulli trials), *p* (probability of success) in cells **A2** and **B2**.
Enter: Formula **=1-B2** in cell **C2**.
Enter: Value **0** in cell **A5**
Enter: Formula **=A5+1** in cell **A6**
Copy: Contents of cell **A6** to cells **A7:A13**
Enter: Formula **=COMBIN($A$2,A5)*$B$2^A5*$C$2^($A$2-A5)** in cell **B5**.

	A	B	C	D	E
1	n	p	q		
2	8	0.5	0.5		
3					
4	x	Pr(x successes)			
5	0	=COMBIN($A$2,A5)*$B$2^A5*$C$2^($A$2-A5)			
6	1				
7	2				
8	3				
9	4				
10	5				
11	6				
12	7				
13	8				

Copy: Contents of cell **B5** to cells **B6:B13**

	A	B	C
1	n	p	q
2	8	0.5	0.5
3			
4	x	Pr(x successes)	
5	0	0.00390625	
6	1	0.03125	
7	2	0.109375	
8	3	0.21875	
9	4	0.2734375	
10	5	0.21875	
11	6	0.109375	
12	7	0.03125	
13	8	0.00390625	

Select: Cells **B5:B13**

Select: On the **Insert** tab, select **Column** on the **Charts** menu. Then, select **2-D (Clustered) Column**.

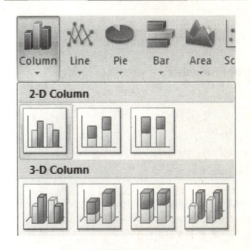

Select: **Design** tab under the **Chart Tools** menu.

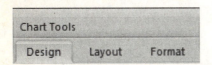

Select: **Select Data** on the **Data** submenu. In the **Select Data Source** window,

Click: **Edit** button under **Horizontal (Category) Axis Labels**

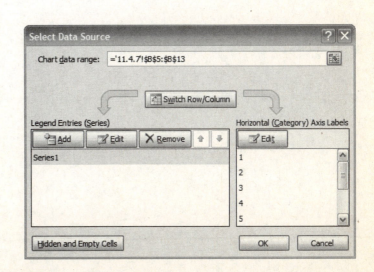

In the **Axis Labels** window,
Select: Cells **A5:A13** <u>using the mouse</u>.
Click: **OK**

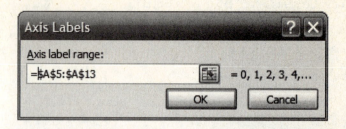

In the **Select Data Source** window,
Select: **OK**

Format chart as desired. For example,
Select: **Layout** tab under the **Chart Tools**
menu.

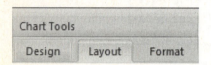

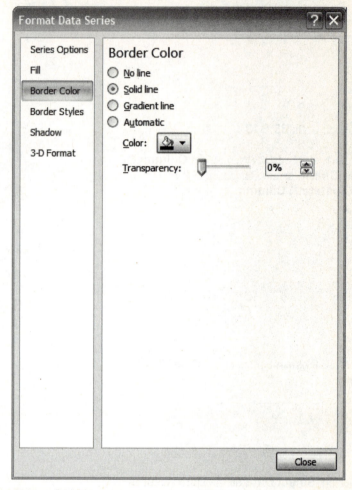

Select: **Series 1** on the **Current Selection**
menu.
Select: **Format Selection**

In the **Format Data Series** window,
Select: **Series Options**
Select: **Gap Width: No Gap**

In the **Format Data Series** window,
Select: **Border Color: Solid line**
Click: **Close**

Select: **Layout** tab under the **Chart Tools** menu.

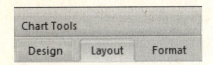

Select: **Data Labels** on the **Labels submenu**
Select: **More Data Label Options**

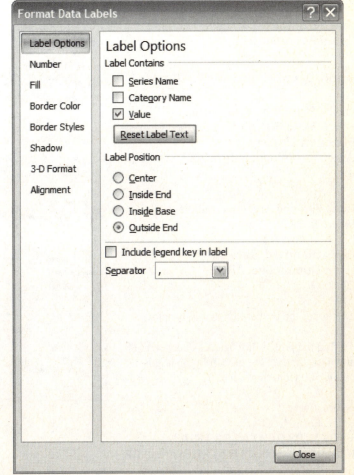

In the **Format Data Labels** window,
Select: **Label Options**
Select: **Label Contains: Value**
Select: **Label Position: Outside End**

In the **Format Data Labels** window,
Select: **Number**
Select: **Number: 3** decimal places
Click: **Close**

Sample output is shown at the right.

To compute the mean (i.e. *np*) and standard deviation (i.e. $\sqrt{npq}$) of this distribution, use the formulas **=A2*B2** and **=SQRT(A2*B2*C2)** respectively.

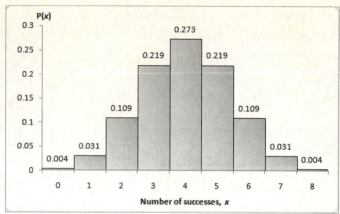

Section 11-5 Normal Distribution

Example 1: Finding Probabilities for a Normal Distribution

Enter: Value **500** (the mean) in cell **A2**
Enter: Value **100** (the standard deviation) in cell **B2**

	A	B	C
1	Mean	Std. Dev.	Prob.
2	500	100	

Select: Cell **C2**
Select: **Formulas** tab

On the **Function Libray** menu,
Select: **More Functions** from the **Function Library**
Select: **Statistical: NORMDIST**

In the Function Arguments window,
Enter: **X: 670**
Enter: **Mean: A2**
Enter: **Standard_dev: B2**
Enter: **Cumulative: TRUE**
Click: **OK**

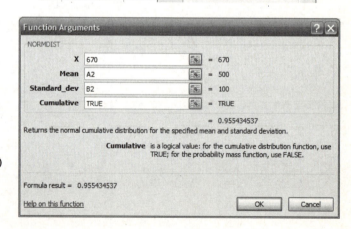

This gives the probability that X is below 670.
To find the probability that X is between 500 and 670, subtract the probability that X is below 500 from this value.

Alternatively,
Enter: Formula
=NORMDIST(670,A2,B2,TRUE)-NORMDIST(500,A2,B2,TRUE) in cell **C2**

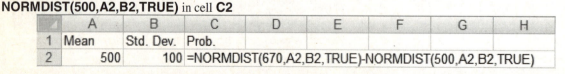

	A	B	C	D	E	F	G	H
1	Mean	Std. Dev.	Prob.					
2	500	100	=NORMDIST(670,A2,B2,TRUE)-NORMDIST(500,A2,B2,TRUE)					

Example 3: Market Research

Enter: Values of *n* (number of repeated Bernoulli trials), *p* (probability of success) in cells **A2** and **B2.**

Enter: Formula **=1-B2** in cell **C2.**

Enter: Formula **=A2*B2** (i.e. *np*) in cell **A5**

Enter: Formula **=SQRT(A2*B2*C2)** (i.e. $\sqrt{npq}$) in cell **B5**

	A	B	C
1	n	p	q
2	20	0.4	0.6
3			
4	Mean	Std. Dev.	Prob.
5		8	=SQRT(A2*B2*C2)

Enter: Formula **=NORMDIST(12.5,A5,B5,TRUE)-NORMDIST(5.5,A5,B5,TRUE)** in cell **C5**

	A	B	C	D	E	F	G	H
1	n	p	q					
2	20	0.4	0.6					
3								
4	Mean	Std. Dev.	Prob.					
5	8	2.19089	=NORMDIST(12.5,A5,B5,TRUE)-NORMDIST(5.5,A5,B5,TRUE)					

To approximate the probability that the sample contains fewer than 4 users of the credit card,

Enter: Formula **=NORMDIST(3.5,A5,B5,TRUE)** in cell **C5**

	A	B	C	D	E
1	n	p	q		
2	20	0.4	0.6		
3					
4	Mean	Std. Dev.	Prob.		
5	8	2.19089	=NORMDIST(3.5,A5,B5,TRUE)		